L. Mark Berliner
Douglas Nychka
Timothy Hoar (Editors)

Studies in the Atmospheric Sciences

Springer

L. Mark Berliner
Department of Statistics
Ohio State University
Columbus, OH 43210

Douglas Nychka
National Center for Atmospheric Research
P.O. Box 3000
Boulder, CO 80307-3000

Timothy Hoar
National Center for Atmospheric Research
P.O. Box 3000
Boulder, CO 80307-3000

Library of Congress Cataloging-in-Publication Data

Studies in the atmospheric sciences / editors, L. Mark Berliner, Douglas Nychka, Timothy Hoar.
    p. cm. -- (Lecture notes in statistics ; 144)
    Includes bibliographical references and index.
    ISBN-13:978-0-387-98757-6   e-ISBN-13:978-1-4612-2112-8
    DOI: 10.1007/978-1-4612-2112-8

    1. Atmospheric physics--Mathematical models. 2. Atmospheric physics--Statistical
methods. 3. Geophysics--Mathematical models. I. Berliner, L. Mark. II. Nychka,
Douglas. III. Hoar, Timothy. IV. Lecture notes in statistics (Springer-Verlag) ; v. 144.

QC861.2 .S78 2000
551.5--dc21

                    99-052678

Printed on acid-free paper.

Camera ready copy provided by the editors.

9 8 7 6 5 4 3 2 1

ISBN-13:978-0-387-98757-6   SPIN 10710835

Preface

The National Center for Atmospheric Research (NCAR) in Boulder, Colorado, is a major institution performing research in the environmental sciences. After an open competition, the National Science Foundation gave its financial support to NCAR for the establishment of a much-needed statistics/probability program in the geophysical sciences. The result is the NCAR Geophysical Statistics Project (GSP). The primary mission of this program is:

- To foster excellence in the use of modern statistical science in the geophysical and environmental sciences.
- To attract and support statistical scientists for collaborative research.

To fulfill its mission, GSP engages in research in statistical science and its application to the atmospheric and allied sciences. It supports a variety of collaborative efforts between statistical scientists and disciplinary scientists. Senior statisticians, both members of and visitors to GSP, work in conjunction with researchers at NCAR and elsewhere in the application of current statistical methods as well as in the development of new statistical models and techniques. Senior GSP members and NCAR researchers also collaborate in the mentoring of junior members, typically post-doctoral level or statistics graduate students.

Mark Berliner was appointed GSP Project Leader in June 1995; Doug Nychka was appointed Project Leader in the Summer of 1997. Since its inception in 1994, GSP has involved numerous senior and junior statisticians in its activities. As of this date, 12 fresh PhD scientists have been members of the Project. Project members have participated in a large variety of collaborations with scientists at NCAR and elsewhere. GSP has evolved into a thriving and permanent section of NCAR. With renewed joint support from both NCAR and the National Science Foundation, we can look forward to an outstanding future.

This book presents some of the research efforts of GSP members. Beyond the first, introductory chapter, the lead authors of the nine remaining chapters are or were GSP post-doctoral level visiting scientists. Some of their chapters offer reviews of statistical approaches to problems arising in the climate and weather sciences. Others are more specialized case studies. Though these chapters are far from complete representations of all GSP activities, we believe that they comprise a solid introduction to the problems encountered in atmospheric sciences, as well as indications of the scope of collaborations between geophysical scientists and statisticians. Our intent is that the book be of value to both environmental scientists and statisticians. Indeed, we believe that the balance between review and specific example presented in this book makes it a good focus or supplement for graduate courses in either statistics or disciplinary programs.

This Preface is an opportunity for us to recognize and thank a number of people who contributed to the success of GSP. First, the original Principal Investigators for the first NSF award, Rol Madden and Rick Katz, started GSP and provided all GSP members with their support and wisdom. Numerous NCAR scientists besides Rol and Rick spent substantial time and effort in initiating collaborations and educating and guiding both the junior and senior members of GSP. In addition to NCAR scientists appearing as coauthors of chapters in this book, several people who have made significant, long-term investments in GSP include Grant Branstator, Ralph Milliff, Kevin Trenberth, Joe Tribbia, Dave Schimel, Dennis Shea, and Tom Wigley. We have also benefitted from the support of the NCAR senior administration, and we particularly mention the help of Robert Serafin, Director of NCAR, Maurice Blackmon, Director of the Climate and Global Dynamics Division, and various past and current NCAR division directors. Finally, Tim Hoar and Elizabeth Rothney have been invaluable stalwarts to the project; the Editors are very grateful to Ms. Rothney for her efforts in completing this volume.

We gratefully acknowledge support from the National Science Foundation under Grants DMS 9815344 and DMS 9312686 for the Geophysical Statistics Project and its research.

L. Mark Berliner
Ohio State University

Douglas W. Nychka
National Center for Atmospheric Research

Lecture Notes in Statistics 144

Edited by P. Bickel, P. Diggle, S. Fienberg, K. Krickeberg,
I. Olkin, N. Wermuth, S. Zeger

Springer
*New York
Berlin
Heidelberg
Barcelona
Hong Kong
London
Milan
Paris
Singapore
Tokyo*

Contents

viii

7 **Neural Networks: Cloud Parameterizations**   97
Barbara A. Bailey, L. Mark Berliner, William Collins,
Douglas W. Nychka, Jeffrey T. Kiehl

8 **Exploratory Statistical Analysis of Tropical Oceanic**
  Convection Using Discrete Wavelet Transforms   117
Philippe Naveau, Mitchell Moncrieff, Jun-Ichi Yano, Xiaoqing Wu

9 **Predicting Clear-Air Turbulence**   133
Claudia Tebaldi, Doug Nychka, Barbara Brown, Bob Sharman

Introduction

L. Mark Berliner
Ohio State University, Columbus, OH 43210, USA

Douglas A. Nychka
National Center for Atmospheric Research, Boulder, CO 80307, USA

Timothy J. Hoar
National Center for Atmospheric Research, Boulder, CO 80307, USA

1 Statistics in the Climate and Weather Sciences

> Global satellite observations coupled with increasingly sophisti-
> cated computer simulations have led to rapid advances in our un-
> derstanding of the atmosphere. While opening many doors, these
> tools have also raised new and increasingly complex questions about
> atmospheric behavior. At the same time, environmental issues have
> brought atmospheric science to the center of science and technol-
> ogy, where it now plays a key role in shaping national and interna-
> tional policy. *Murray Salby, 1996*

Even a cursory attention to current events confirms the profound influence
of short-term weather and long-term climatic change on our Earth's inhab-
itants. Weather prediction plays a significant role in the planning of human
affairs. Safety of our lives and property relies on the tracking and prediction
of weather events such as hurricanes/typhoons, severe thunderstorms, torna-
does, blizzards, floods, and atmospheric turbulence. A variety of scientific,
political, and economic concerns, and now the general public, are focused on
the behavior of El Niño and La Niña (or, more precisely, the El Niño–Southern
Oscillation, ENSO, phenomenon). Further, a broader appreciation of the role
of weather and climate impacts on the environment of the planet has now
led to nearly universal concern regarding potential climate change, its causes,
impacts, and possible remediation.

There exists substantial physical understanding of fundamental physics controlling our atmosphere. The Earth's atmosphere and oceans are *fluids*. Their
behaviors depend on the energy input from the Sun, subsequent exchange and
radiation of energy, and are moderated by gravity and the rotation of the
Earth. In principle, the laws of physics yield a collection of partial differential
equations whose solution would provide a prescription for the evolution of the

fluids, depending on initial and boundary conditions. It may seem then that all we need to do is provide the requisite initial and boundary conditions. However, there are serious limiting factors. The fundamental equations are complex nonlinear systems. Indeed, we do not anticipate being able to solve these equations explicitly even with current advances in computing. Further, the specific form of nonlinearity present leads to *chaos*: even the slightest uncertainties in initial and boundary conditions eventually lead to major uncertainties in solutions, uncertainties so large that the value of any specific solution is negligible. The role of chaos in the weather was anticipated by H. Poincaré, though discovered and studied by E. Lorenz in the 1960s. Lorenz' work was critical to the modern view in the atmospheric sciences and led to significant interest in other disciplines. (See Lorenz [Lor93] for discussion; Berliner [Ber92] offers a general review for statisticians.)

The massiveness of the system modeled poses other problems. Approaches to obtaining solutions to the equations involve a number of approximations. First, solutions are computed numerically, and, hence, subject to numerical error. Second, the computer representations of the equations are typically based on discretizations in both time and space. This means that there are important physical processes that are not *resolved* by the numerical model. For example, clouds play a critical role in the atmosphere, but are typically much smaller than model gridboxes. Further, the physics needed to understand the evolutions of the atmosphere or ocean involve many interactions including exchanges at the boundary between atmosphere and ocean. Finally, not all relevant physical variables can be included in the model.

In the face of such complicated sources of uncertainty, combined with the increasing size of observational datasets, we believe that the problems of weather and climate are best stated and answered in the language of statistics. Indeed, one might claim that the Earth's climate system is a stochastic process; hence, we should abandon the strong use of physics in favor of statistical prediction models. However, this perspective also has problems. First, and most obvious, the extraction of information from massive datasets is very difficult. A more critical issue is that, though large, these datasets are often sparse relative to the geophysical phenomenon of interest. Further, the amount of data available varies with the subsystem considered. Compared to atmospheric data, the observational data available about the Earth's oceans are a mere trickle. The complexity of the Earth's climate system that limits the success of purely physical approaches also limits purely statistical approaches. Our view is that ultimate success will involve the combination of viewpoints; a synthesis between statistics and physics.

A large variety of statistical methods are used routinely in the atmospheric sciences. For example, techniques of multivariate time series are especially common. These include multivariate autoregressive, moving average models and Kalman filtering. Statistical methods for spatial data are also standard.

For example, *Kriging* (often called *objective analysis* or *optimal interpolation*) plays a crucial role in the study of spatial *fields*. A major tool in the analysis of space–time data is *empirical orthogonal functions* (EOF). Statisticians will recognize EOF analysis as a relative of principal components and canonical correlation analysis. See Cherry [Che96b] for further discussion and review. Stochastic modeling is common in many investigations. Many climate scientists use stochastic processes theory, including topics such as stochastic differential equations and Markov processes. Authoritative reviews and further reference may be found in [Wil95] and [von98].

However, advances in computational assets have led to new approaches in statistical analyses. Computer intensive data analyses include hybrid estimation procedures (e.g., neural nets, nonparametric regression and density estimation, wavelet-based data analysis, EM algorithm, etc.). Computational advances have also led to a revitalization of the Bayesian approach to data analysis. These topics are not found in the classical introductory texts in statistics and we believe this volume fills a need for presenting these new tools.

2   A Guide to this Volume

2.1   Chapter Outline

Each chapter pairs an area of application in the atmospheric sciences with innovative statistical methodology. Chapters 2 through 5 can be considered as reviews, tending to give broad surveys of selected statistical methods. The remaining chapters, 6–10, are case studies, tackling specific scientific problems and bringing to bear the appropriate statistics and sometimes developing new methods in the process. Although these later chapters track specific applications, the topics have been chosen to represent a suite of modern statistical tools for the geosciences. Each of these chapters will serve to introduce the reader to these methods.

2.1.1   Forecasting

Chapter 2 leads this volume with statistical foundations of modern weather forecasting. This involves combining meteorological observations with an estimate of the current state of the atmosphere to produce a better estimate. In this context, Bayes' theorem is a basic starting point to think about forecasting. The introduction of some new results on ensemble forecasting and non-Gaussian models also provides a good base for the subsequent chapters.

2.1.2   Spatial and Temporal Processes

The next three chapters are about spatial and space–time models for geophysical processes. Chapter 3 on spatial models contrasts the traditional geostatistics approach to modeling covariances with a more direct regression approach.

A general theme in this chapter and the next is the use of hierarchical models to generate spatial dependence in a way that is physically meaningful. Nesting these within a Bayesian framework also facilitates including physical prior knowledge. In the past ten years we have witnessed a watershed of Bayesian applications largely due to the ability to sample complicated posterior distributions by Markov chain Monte Carlo (MCMC). The geophysical applications in these two chapters, estimating temperature and wind fields, would not have been feasible without MCMC techniques. The last member of this group (Chapter 5) returns to the forecast problem but concentrates on spatial design. The area of targeting observations in areas where the data will "best" improve forecasts is an active area of meteorological research. Casting the targeting question as a statistical design problem provides a framework for clarifying the optimization criteria, comparing various suggestions for targeting, and developing new procedures.

2.1.3   Functional Data

In many cases the basic observational data can be organized in chunks where each chunk is points sampled from a curve. The motivating application in Chapter 6 is the daily concentration curve of atmospheric ozone at a specific location as a function of height. Here the chunks are each day's ozonesonde, being a vertical profile for ozone. In order to understand how the distribution of ozone changes over time. Readers will recognize some standard methods, such as principle component analysis and regression, coupled with more specialized techniques such as spline interpolation. Taken together they provide a comprehensive tool for modeling the shape of ozone profiles.

2.1.4   Neural Networks

Often regression problems involve nonlinear relationships and large numbers of potential covariates. Neural network models are useful in cases where one cannot specify a functional form for the regression model, and when one is uncertain which covariates are important to include in the regression analysis. The application in Chapter 7 is to build a model for the cloud amount on a grid of spatial locations based on the past cloud amounts and covariates for cloud formation. Although the functional form for neural networks may seem exotic at first, they can be fitted by nonlinear least squares. Results of this case study provide accurate representations of the spatial cloud process.

2.1.5   Wavelets

Geophysical processes often have different properties at different scales. In Chapter 8, the structure of thunderstorms (strong convection) is analyzed at different spatial scales by a wavelet analysis. A property of a wavelet decomposition is to divide a spatial image into horizontal, vertical, and diagonal features of different sizes. This makes it possible to find linear patterns in an image and helps to distinguish among different types of convection. A distinctive feature of this case study is that the datasets are actually simulation results from a deterministic numerical model. Nevertheless, a statistical analysis is appropriate in order to summarize basic features of the model output.

2.1.6   Flexible Classification

Severe weather and small-scale phenomenon are difficult to forecast directly using numerical weather models. Chapter 9 relies on statistical methods to forecast the presence or absence of clear-air turbulence (CAT) that affects aircraft. Although the covariates used for the forecast model are derived from a numerical weather prediction model, these are used in a statistical discriminant analysis to classify regions of the United States for their tendency to have CAT. The classification methodology is a powerful combination of conventional, linear discriminant analysis and more recent advances in nonparametric regression and model selection (multivariate adaptive regression splines). The net results are forecast models that perform better than past indices for turbulence and can adapt to changing meteorological conditions.

2.1.7   Local Variograms

The last case study in this book is a spatial data analysis that aids in understanding the spatial scales of ocean processes. In particular, Chapter 10 quantifies the extent of the spatial correlation of phytoplankton in the Atlantic Ocean based on remotely sensed amounts of chlorophyll. This is an extensive use of variograms to analyze spatial data. Although the variogram is an established technique in geostatistics, its application to spatial subsets of a dataset to see how spatial structure changes is not as common. In this case, the estimates of correlation lengths confirm patterns of ocean circulation derived from physical models. Although this case study addresses scientific issues in the ocean, the methods can be readily transferred to atmospheric spatial data.

2.2   What Is Missing?

2.2.1   Statistical Methods

Some areas of geostatistics have been omitted from this volume. One important area is times series methods to detect trends or other signals in data and to characterize the power of a time series at different frequencies. Some of these statistical tools are used in detecting climate change (see [Lev99]) and so should not be ignored. Another omitted topic is empirical orthogonal functions (EOFs), used to represent a spatial/temporal data through a decomposition into principle components and already mentioned in the first section of this chapter. Chapter 6 does make use of EOFs as an intermediate step in the analysis of ozone profiles but does not emphasize the ubiquity of EOF analyses in the atmospheric science literature.

2.2.2 Scope

The separation of the geosciences into distinct disciplines such as meteorology, oceanography, and ecology fails to emphasize the connections between the Earth's physical and biological systems. For example, because of the interaction of the ocean and atmosphere at their boundary layers, it is often artificial to be restricted just to atmospheric processes. As a vehicle for introducing innovative statistical methods into the geosciences we felt that a focus on the atmosphere was the right scope for a single volume. However, the statistical methods covered here translate easily into other scientific contexts. For example, we believe that Bayesian approaches based on hierarchical modeling will have an impact for understanding ocean and biogeochemistry processes as well as the atmosphere. Thus, we expect that the applications in this volume provide broader statistical support for studying the Earth's systems and the human influences on the environment.

3 Software, Datasets, and the Web Companion

In order to provide a flexible and adaptive way to make software and datasets available to readers, we have chosen to deemphasize these topics in the printed version of this volume and rely on a digital companion to fill this need. We will refer to this electronic source of software, data, and updated references as the *web companion*. At the time of publication the homepage for this resource can be accessed from the homepage of the Geophysical Statistics Project at the National Center for Atmospheric Research and has the address: http://www.cgd.ucar.edu/stats.

A Statistical Perspective on Data Assimilation in Numerical Models

Gary Sneddon
Memorial University of Newfoundland
St. John's, Newfoundland, CANADA

1 Introduction

The phrase *data assimilation* is used in a variety of contexts in ocean and atmospheric studies, and the geosciences in general. Broadly speaking, data assimilation involves combining observational data with numerical results to yield a better prediction. It often has the sound of a "black-box procedure," where one puts in data at one end and it becomes "assimilated" as it comes out of the other. However, this is certainly not the case. One of our goals in this chapter will be to discuss the nature of data assimilation, in particular to define it in a general sense, and to draw out many of the ideas in assimilation methods that are familiar to statisticians, but phrased in different ways. In this way we hope to fulfill one of the objectives of this volume; to bridge gaps and to find common ground between methods in statistics and the geosciences.

Before the advent of extensive observational networks and technological advances that yield large amounts of data, geoscientists relied on subjective analysis [Dal91, p. 17] to represent observations on maps. This was done with the hope that the physical laws governing atmospheric flows could be derived from the maps. Objective analysis procedures [Dal91, pp. 21] followed, in order to estimate the dependent variables on a regular grid from scattered observations. This led to the technique of statistical interpolation [Gan65]. Methods have continued to move forward, with data assimilation now an increasingly popular method of estimation.

Daley [Dal91, pp. 11–12] discusses sources of data for data assimilation and places them in three general categories based on the instruments used for collection:

1. point instruments such as thermometers;
2. remote instruments that sample an area or volume (of atmosphere); and
3. instruments that calculate wind velocities from Lagrangian trajectories[1]

The characteristics of these instruments need to be understood in order to assimilate their information in an optimal way. Talagrand [Tal97] also discusses observations as being direct or indirect measurements, and how they

---

[1] An object is followed remotely and velocities are found from observed displacements at fixed times.

can vary significantly in nature and accuracy. These properties should also be incorporated in how the data is assimilated.

To a statistician, the amount of data available in some geophysical problems often appears to be tremendous: hourly temperature readings over a region, daily precipitation amounts for cities in the United States, and so on. For the statistician used to working with modest size datasets, this can pose both conceptual and technical challenges. Besides the size of geophysical datasets, two other features characterize assimilation problems. First, since there is a vast amount of data available in some situations, should any estimates we construct simply rely on the data, with no explicit use of any known governing physical process? Second, there are many problems where there appears to be a lot of data, but we are trying to construct estimates on a very fine grid, where the number of gridpoints will exceed the number of observations [Lor86]. To consider a comparable problem in statistics, imagine a regression model with 10 observations and 20 independent variables. This is often referred to as an underdetermined regression model, and we cannot find a unique parameter estimator without imposing some prior knowledge or constraint on the parameters.

We have alluded to three sources of information available: physical laws, observations, and prior information. It seems clear that most of us would not want to simply operate in one of these extreme situations, using only one source of information. Instead we would like to combine our knowledge of the areas in some fashion. This is what data assimilation attempts to do. We can think of data assimilation as a procedure that combines: available data; a model, often derived from equations that govern the physical process; and a priori knowledge of the field we are trying to estimate.

This chapter is structured as follows. In Section 2 we discuss assimilation methods which ignore the time component, and describe how they can be viewed as penalized least squares problems. We then move on to discuss time-dependent assimilation methods, with the focus being on the Kalman filter. In this section we will discuss a recent and promising method of assimilation in operational numerical models, called ensemble forecasting. The chapter concludes with a numerical study which addresses one of the problems that often exists in assimilation methods: implicit assumption of linear models with Gaussian errors.

2   Assimilation and Penalized Least Squares

Our starting point is a model of the following form:

$$\mathbf{y} = \mathbf{h}(\mathbf{x}) + \mathbf{e}, \tag{2.1}$$

where $\mathbf{e}$ is an error term. (We often have $\mathbf{y} = \mathbf{y}_t$, $\mathbf{x} = \mathbf{x}_t$, but we ignore the possible time dependence until later sections.) In general, $\mathbf{h}(\mathbf{x})$ may be a nonlinear function, but we initially linearize the model and rewrite (2.1) as

$$\mathbf{y} = \mathbf{Hx} + \mathbf{e}. \tag{2.2}$$

This is the general linear model commonly used in statistics. We usually want to estimate $\mathbf{x}$, which is often called the *state variable* in geoscience terminology. Now, what do $\mathbf{x}$ and $\mathbf{y}$ represent? In many problems in the Earth sciences, $\mathbf{y}$ is a vector of observations, say temperature readings at weather stations in a particular region. The $\mathbf{x}$ vector may represent temperature or another variable on a regular grid over the region. Model (2.1) may be derived from the physical process, or it may be based on an interpolation method that maps the observations to the gridpoints. Using $\mathbf{y}$ and (2.2) to make inferences about $\mathbf{x}$ is sometimes referred to as an inverse problem, which are dealt with extensively by Tarantola [Tar87].

We stated previously that data assimilation is an estimation procedure that attempts to combine three sources of information to produce an estimate of the state vector. In climate research, and other areas of the geosciences, this is often done by minimizing an expression like the following:

$$J(\mathbf{x}) = \frac{1}{2}(\mathbf{x} - \mathbf{x}_a)^T \mathbf{R}^{-1}(\mathbf{x} - \mathbf{x}_a) + \frac{1}{2}(\mathbf{y} - \mathbf{Hx})^T \mathbf{P}^{-1}(\mathbf{y} - \mathbf{Hx}), \tag{2.3}$$

where $\mathbf{x}_a$ is an estimate from a previous forecast, and $\mathbf{R}$ and $\mathbf{P}$ are the forecast and observation error covariance matrices, respectively. In many disciplines minimizing (2.3) is called a variational problem.

A statistician will recognize the problem posed in (2.3) as a penalized least squares problem, where we are constraining the solution via the first term in (2.3). The estimate $\hat{\mathbf{x}}$ which minimizes (2.3) is easily found to be

$$\hat{\mathbf{x}} = (\mathbf{H}^T \mathbf{P}^{-1} \mathbf{H} + \mathbf{R}^{-1})^{-1}(\mathbf{H}^T \mathbf{P}^{-1} \mathbf{y} + \mathbf{R}^{-1} \mathbf{x}_a). \tag{2.4}$$

Variants of (2.3) also appear in work in spline modeling [Wah90] and density estimation [Sim96]. In these situations we wish to minimize the misfit between the observations and a functional form, subject to a smoothness constraint on the function:

$$\sum_{i=1}^{n} [y_i - f(x_i)]^2 + \lambda \int [f^{(k)}(x)]^2 \, dx.$$

Typically, $k = 2$ is chosen to penalize curvature, and λ is thought of as the smoothing parameter.

An alternative derivation of (2.3) is from a Bayesian approach. Assume that $\mathbf{e} \sim N(\mathbf{0}, \mathbf{P})$ and the prior distribution for $\mathbf{x}$, $\pi(\mathbf{x})$, is $N(\mathbf{x}_a, \mathbf{R})$. Then our inference on $\mathbf{x}$ should be based on the posterior distribution of $\mathbf{x}$, given $\mathbf{y}$, which is

$$\pi(\mathbf{x}|\mathbf{y}) = \frac{f(\mathbf{y}|\mathbf{x})\pi(\mathbf{x})}{f(\mathbf{y})} \tag{2.5}$$
$$\propto f(\mathbf{y}|\mathbf{x})\pi(\mathbf{x}).$$

Under the Gaussian assumptions we find

$$\pi(\mathbf{x}|\mathbf{y}) \propto \exp\left[-\frac{1}{2}(\mathbf{x}-\mathbf{x}_a)^T\mathbf{R}^{-1}(\mathbf{x}-\mathbf{x}_a) - \frac{1}{2}(\mathbf{y}-\mathbf{H}\mathbf{x})^T\mathbf{P}^{-1}(\mathbf{y}-\mathbf{H}\mathbf{x})\right]$$

$$= \exp\left[-\frac{1}{2}\left(\mathbf{x}^T(\mathbf{H}^T\mathbf{P}^{-1}\mathbf{H}^T + \mathbf{R}^{-1})\mathbf{x} - 2(\mathbf{y}^T\mathbf{P}^{-1}\mathbf{H} + \mathbf{x}_a\mathbf{R}^{-1})\mathbf{x}\right)\right.$$

$$\left. - \frac{1}{2}\left(\mathbf{x}^T\mathbf{P}^{-1}\mathbf{x} + \mathbf{x}_a^T\mathbf{R}^{-1}\mathbf{x}_a\right)\right].$$

We can see that $-\log$ (posterior in (2.6)) is a constant plus (2.3). Therefore the penalized likelihood or penalized least squares calculation is just finding the posterior mode.

Suppose we now define

$$\mathbf{A} = \mathbf{H}^T\mathbf{P}^{-1}\mathbf{H} + \mathbf{R}^{-1}, \quad \mathbf{b}^T = \mathbf{y}^T\mathbf{P}^{-1}\mathbf{H} + \mathbf{x}_a^T\mathbf{R}^{-1}, \quad c = \mathbf{y}^T\mathbf{P}^{-1}\mathbf{y} + \ddot{\mathbf{x}}_a\mathbf{R}^{-1}\mathbf{x}_a.$$

Then we can rewrite the expression as

$$\pi(\mathbf{x}|\mathbf{y}) \propto \exp\left[-\frac{1}{2}\left((\mathbf{x}_a - \mathbf{A}^{-1}\mathbf{b})^T\mathbf{A}(\mathbf{x}_a - \mathbf{A}^{-1}\mathbf{b}) - \mathbf{b}^T\mathbf{A}^{-1}\mathbf{b} + c\right)\right].$$

The final terms do not depend on $\mathbf{x}$, and so the posterior distribution is Gaussian, with mean

$$\mathbf{A}^{-1}\mathbf{b} = (\mathbf{H}^T\mathbf{P}^{-1}\mathbf{H} + \mathbf{R}^{-1})^{-1}(\mathbf{H}^T\mathbf{P}^{-1}\mathbf{y} + \mathbf{R}^{-1}\mathbf{x}_a)$$

and covariance

$$\mathbf{A}^{-1} = (\mathbf{H}^T\mathbf{P}^{-1}\mathbf{H} + \mathbf{R}^{-1})^{-1}.$$

We see that the posterior mean is $\hat{\mathbf{x}}$ in (2.4). Therefore the cost function (2.3), sometimes used as a first principle in geophysical work, can be justified as a Bayesian method based on multinormal distributions for the observations and the prior.

Although the value of $\hat{\mathbf{x}}$ has a simple form, it is not always a simple estimate to calculate. Problems in the atmospheric sciences can be very large, with upward of 10^5 state variables to estimate. This means that the size of $\mathbf{R}$ in (2.4) would be $10^5 \times 10^5$, and we need its inverse. That calculation is computationally intensive, so the evaluation of the formula in (2.4) is very difficult, if not impossible, unless the covariance matrices have some simple, sparse structure. Daley [Dal97] states that the forecast error correlations are often assumed to be stationary, homogeneous, and isotropic, which aids in finding a solution. However, it also is an assumption that is unrealistic.

Finally, the derivation of (2.4) relies on (2.2) being correct, or at least a good approximation to (2.1). This may not always be the case. The formulation also ignores any model error that may be present. Instead we just have an error term associated with the observations. We have also mentioned the issue of non-Gaussian errors, and that they will sometimes be inappropriate. One example would be precipitation data, which is never nonnegative, and has the positive probability of being zero.

3 Time-Dependent Assimilation Methods

Our discussion on penalized least squares did not take explicit account of the temporal component that will be present in many geoscience problems, particularly those which deal with the ocean or atmosphere. For example, in *numerical weather prediction* (NWP) one is concerned with forecasting the atmospheric state at time t, given all the previous information up to and including time $t-1$. Then, when data become available at time t, we would update our estimate to use this extra information.

This formulation fits naturally into the context of the Kalman filter. The Kalman filter has its origins in engineering (see the seminal works of [Kal60a] and [Kal60b]), especially signal processing [Jaz70]. It has a wide range of applications, including time series analysis [Shu88, pp. 173–184]. Since the Kalman filter constructs estimates sequentially in time, it has appeal for use in atmospheric science models. We will also see that the problem formulation for the Kalman filter falls under our description of an assimilation method.

3.1 The Kalman Filter

We begin by phrasing our problem as a linear state–space model:

$$\mathbf{y}_t = \mathbf{H}_t \mathbf{x}_t + \mathbf{e}_t, \tag{2.6}$$
$$\mathbf{x}_t = \mathbf{W}_t \mathbf{x}_{t-1} + \mathbf{v}_t. \tag{2.7}$$

Here $\mathbf{y}_t$ is an $n \times 1$ vector of observations at time t and $\mathbf{x}_t$ is the $p \times 1$ state vector we wish to estimate at time t. We refer to (2.6) as the measurement equation and (2.7) as the state equation. We assume that $\mathbf{e}_t \sim \mathrm{N}(\mathbf{0}, \boldsymbol{\Sigma}_e)$, $\mathbf{v}_t \sim \mathrm{N}(\mathbf{0}, \boldsymbol{\Sigma}_v)$, and that $\mathbf{e}_t$ and $\mathbf{v}_t$ are independent.

The structure of the state–space model is a natural one for problems in the atmospheric sciences. The observation equation (2.6) maps the state vector to the observations. The state equation (2.7) has its roots in the physical equations that describe the evolution of the atmosphere. The matrix $\mathbf{W}_t$ is typically based on a linearized and discretized version of these equations, and a stochastic term is included to allow for uncertainty in the model.

Let $\mathbf{x}_{t-1}^a$ and $\mathbf{P}_{t-1}^a$ denote the Kalman filter estimate and associated covariance of the state $\mathbf{x}_{t-1}$ based on all observations up through time $t-1$. The Kalman filter predictor ($\mathbf{x}_t^f$) and associated covariance ($\mathbf{P}_t^f$) for forecasting $\mathbf{x}_t$ based on data up through time $t-1$ are

$$\mathbf{x}_t^f = \mathbf{W}_t \mathbf{x}_{t-1}^a, \tag{2.8}$$
$$\mathbf{P}_t^f = \mathbf{W}_t \mathbf{P}_{t-1}^a \mathbf{W}_t^T + \boldsymbol{\Sigma}_v, \tag{2.9}$$

while the analysis estimate and estimated covariance of $\mathbf{x}_t$ based on data up through time t are

$$\mathbf{x}_t^a = \mathbf{x}_t^f + \mathbf{K}_t(\mathbf{y}_t - \mathbf{H}_t \mathbf{x}_t^f), \tag{2.10}$$
$$\mathbf{P}_t^a = (\mathbf{I} - \mathbf{K}_t \mathbf{H}_t)\mathbf{P}_t^f, \tag{2.11}$$

where

$$\mathbf{K}_t = \mathbf{P}_t^f \mathbf{H}_t^T (\mathbf{H}_t \mathbf{P}_t^f \mathbf{H}_t^T + \mathbf{\Sigma}_e)^{-1}$$

is the Kalman gain matrix.

There are a number of ways to derive the Kalman filter estimators. Some rely on explicit assumptions of Gaussian errors in the state–space model; see [Coh97] for one approach. They can also be derived by relying only on the first and second moments of the distributions being specified [Kal60b]. The estimators are then the best linear unbiased predictors.

Perhaps a more natural and general derivation is from a Bayesian formulation of the problem, as discussed by [Wes97]. Given the state–space model (2.6)–(2.7), and a specified prior on the initial information $\mathbf{x}_0$, we find (2.8) and (2.9) are the mean and covariance of the prior distribution of $\mathbf{x}_t$, given all information available at time $t-1$, while (2.10) and (2.11) are the mean and covariance of the posterior distribution of $\mathbf{x}_t$ given all information available at time t. Therefore, as in the penalized least squares work, we can view the classical estimators as Bayes estimators. This standard Bayesian calculation, as explained by [Wes97], can be applied in principle in all non-Gaussian models to derive the posterior distribution of $\mathbf{x}_t$. We will return to these various methods later in this section.

Although the Kalman filter has a simple formulation, there are practical problems in using it in atmospheric science models, particularly in an operational setting. First, the evolution of the state vector is usually a nonlinear function

$$\mathbf{x}_t = \mathbf{w}_t(\mathbf{x}_t) + \mathbf{v}_t, \qquad (2.12)$$

arising from the physical equations that describe the evolution of the atmosphere. Therefore our estimators in (2.8)–(2.11) cannot be used in their given forms. One way around this is to use a linear approximation to $\mathbf{w}_t$, and use the *extended* Kalman filter [Coh97]. The Bayesian calculations can also be extended to utilize linearizations in (2.8)–(2.11) [Wes97].

Another issue is the dimension of the covariance matrices, as we discussed in Section 2. The problem is magnified in the Kalman filter because we need to invert the forecast and analysis covariance matrices at each timestep of the procedure. A great deal of work has gone into the specification and estimation of these covariance matrices. Fisher and Courtier [Fis95] discuss several methods of estimating the analysis covariance matrix. The methods usually rely on being able to find some of the eigenvalues and eigenvectors of the relevant matrices, without requiring explicit knowledge of the entire matrix. These eigenvectors can then be used to construct low-rank approximations to the covariance matrices.

We mentioned various methods can be used to derive the Kalman filter estimators. If the error distributions are Gaussian (or nearly Gaussian), then the estimators are optimal for quadratic prediction error. However, when the distributions depart from Gaussian, the Kalman filter estimators may be very poor. Cohn [Coh97] discusses the assumption of non-Gaussian errors and the incorrect analyses that can result if there is a mismatch between the true error structure and what is assumed in the assimilation method. We again point out that Bayes' theorem can be used as a general approach to derive the estimators when the distributions are non-Gaussian.

4   Ensemble Forecasting

The advent of powerful computing systems has allowed data assimilation to become a practical method of estimation in many atmospheric science problems. This is particularly true in problems such as weather forecasting, where estimation must be done in real time. It is of little help to have an assimilation method which requires 8 hours of computing time to construct a 3-hour forecast. A similar progression has occurred in statistics, where methods such as the bootstrap, CART modeling, and the Gibbs sampler have become viable methods as computing resources have increased.

One of the more recent assimilation methods being used in NWP is ensemble forecasting. This procedure, while discussed by Leith [Lei74] around 25 years ago, was simply not a feasible option at the time. The intensive computational burden of ensemble forecasting will always be a limiting constraint, but it is becoming a more manageable problem. Ensemble forecasting is not restricted to NWP, but we will focus our attention on this particular example.

4.1   Methodology

In NWP one can think of using a nonlinear state–space model as the general formulation for the problem. Therefore the state vector is evolving forward in time as in (2.12). Mathematically, we begin with an initial condition for the model, denoted x_0, which is used to begin the forecast and give estimates of the form $x_t^f = w_t(x_{t-1}^a)$. This forecast estimate would be updated to an analysis estimate when data become available. This estimation procedure at consecutive time steps is often called the forward integration of the model.

Ensemble forecasting modifies this procedure. Instead of beginning with one set of initial conditions, one begins with a collection (or ensemble) of k initial conditions, denoted $x_{0,1}, \ldots, x_{0,k}$. Each of these ensemble members are integrated forward in time as above, yielding analysis estimates $x_{t,1}^a, \ldots, x_{t,k}^a$ at time t. Therefore at time t we have an ensemble of k estimates. If we take the mean of the ensemble members, we can imagine that we are employing a filter that retains aspects of the forecasts that are common while filtering out those that vary among members. This is particularly useful when the model integration is nonlinear.

There are different methods used to generate the ensemble members. One that is a familiar method to statisticians is a Monte Carlo approach that generates the initial conditions by making an assumption about the distribution on the initial state. For example, we could assume that the initial state has a *probability density function* (pdf) $p(\mathbf{x})$, and we would draw a sample $\mathbf{x}_{0,1}, \ldots, \mathbf{x}_{0,k}$ from $p(\mathbf{x})$ [Ber99a]. A Monte Carlo approach is also discussed by [And99], [Hou98], and [Bui97]. As we will discuss later, the Monte Carlo method can be viewed as approximating a random sample from the pdf of interest.

A slightly different Monte Carlo method is discussed by [Ham99] and [Bur98]. Although they state that they are approximating the evolution of the pdf $p(\mathbf{x})$, they do this by perturbing the observations at each time point. If we assume the state–space model is linear, they create

$$\mathbf{y}_{t,i} = \mathbf{y}_t + \mathbf{e}_i, \qquad i = 1, \ldots, k,$$

where $\mathbf{e}_i \sim N(\mathbf{0}, \boldsymbol{\Sigma}_e)$. The ensemble members are then updated as

$$\mathbf{x}_{t,i}^a = \mathbf{x}_{t,i}^f + \mathbf{K}_t(\mathbf{y}_{t,i} - \mathbf{H}_t\mathbf{x}_{t,i}^f).$$

Toth and Kalnay [Tot97] have developed a second approach, called the breeding method. It can be viewed as a dynamically constrained technique that generates perturbations from a control forecast that will grow or have grown rapidly [Ham99]. This is in contrast to the stochastic formulation in the Monte Carlo approach. In particular, [Tot97] generate these perturbations in directions where past forecast errors have grown most rapidly, where the differences among the forecasts must be renormalized during the model integration.

There are other reasons why ensemble forecasting is an appealing method. The use of the ensemble members may facilitate estimating the forecast and analysis covariance matrices (2.9) and (2.11). It also avoids some of the problems that arise when using the extended Kalman filter if the state–space model is nonlinear, as discussed by [Bur98] and [And99]. One of these gains is a lower numerical cost, since one can often avoid creating a linearized version of the model, and only perform k integrations of the nonlinear model.

4.2 Statistical Perspective

We end by mentioning some relevant statistical aspects in ensemble forecasting. Our goal is to show that ensemble methods are easiest to interpret as a Monte Carlo approximation to a formal Bayesian computation of a posterior distribution. In essence, the method is attempting to generate a sample from the posterior distribution of interest.

To follow our discussion on a Monte Carlo method of ensemble forecasting, suppose that the initial state for our forecast has the pdf $\pi(\mathbf{x}_0)$. For convenience we assume that

$$\pi(\mathbf{x}_0) = N(\boldsymbol{\mu}_0, \mathbf{P}_0).$$

Under the assumption that the state space model (2.6)–(2.7) is valid, what we want to derive is the posterior pdf pdf $\pi(\mathbf{x}_t|\mathbf{y}_t)$. Since we assumed the error distributions in (2.6) and (2.7) are Gaussian, and the model is linear, it is not difficult to write down $\pi(\mathbf{x}_t|\mathbf{y}_t)$ explicitly. However, we mentioned earlier why this cannot be done practically in many atmospheric science problems. Our alternative is to generate a random sample from $\pi(\mathbf{x}_t|\mathbf{y}_t)$.

We begin by generating a random sample from $\pi(\mathbf{x}_0)$, which we again denote by $\mathbf{x}_{0,1}, \dots, \mathbf{x}_{0,k}$. We then take each $\mathbf{x}_{0,i}$ and use the Kalman filter formulas to integrate these vectors forward in time. The $\mathbf{x}_{t,i}^a$ vectors are usually interpreted as a random sample from $\pi(\mathbf{x}_t|\mathbf{y}_t)$, as desired. This sample can then be used to estimate the mean and the covariance of the posterior. However, [Ber99a] points out that these can be poor summaries of the posterior in many situations.

Although the ensemble gives us a random sample from the posterior distribution, we would like a better description of the posterior distribution. One alternative is to use the ensemble to construct a kernel density estimate of the posterior, as discussed by [Ber99a] and [And99]. Typically, this kernel estimator would be of the form

$$\pi(\mathbf{x}_t|\mathbf{y}_t) \approx \sum_{i=1}^{k} \alpha_i \mathrm{N}(\mathbf{x}_{t,i}^a, \mathbf{A}_i), \qquad (2.13)$$

where $\sum_i \alpha_i = 1$ and the $\mathbf{A}_i$ matrices are bandwidth matrices. This idea is appealing, but leads to other issues that must be considered. One of these is that the number of ensemble members is usually much smaller than the dimension of $\mathbf{x}$. Most results in kernel estimation rely on the dimension of $\mathbf{x}$ being much smaller than the number of distributions being used in (2.13).

5 Numerical Studies

We now introduce some numerical results in a problem which examines some of the issues that may not always be adequately addressed in an assimilation procedure. We take a Bayesian approach in studying an example where an assimilation method may miss important features in the estimates.

5.1 Precipitation Model

We will discuss an analysis of a precipitation model that is an extension of work in a one-dimensional model discussed by [Err99].

Assimilation methods used in the geosciences often reduce to a penalized least squares problem, without serious consideration of the error structure being implicitly assumed. The implicit assumptions of approximate linearity and Gaussian errors made in penalized least squares may not always be valid, and it is desirable to have some idea of the consequences of the incorrect assumptions. We will examine this in the context of a limited, but nonlinear model for convective precipitation. The goal of this study is to use a Bayesian analysis to help understand sensible ways of assimilating precipitation data. Although substantial work is required to scale this analysis for operational forecasting, the study still has practical implications. It may suggest appropriate assimilation methods to use and indicate what measures of the goodness-of-fit of the model are sensible when assimilating precipitation observations.

Precipitation is a very localized phenomenon when we consider the scale of a climate model. Therefore the occurrence of precipitation is usually parameterized in some fashion in the model. We consider a Relaxed Arakawa-Schubert (RAS) [Moo92] scheme, which produces a rainfall amount r at a particular location as

$$r = f(\mathbf{x}) = f(\mathbf{T}, \mathbf{q}), \tag{2.14}$$

where $\mathbf{T}$ is a vector of average temperature and $\mathbf{q}$ is a vector of average specific humidity readings at different levels of the atmosphere. The number of levels used in (2.14) can vary; our results will use an eight-level model. Thus the size of our state vector $\mathbf{x}$ will be $2 \times 8 = 16$. Our goal is to make statements about the state vector using a single precipitation observation at the surface. The specific assimilation problem is to improve our prior estimate of $\mathbf{x}$ by incorporating the observed precipitation value.

We assume that there is error present in the model and the observation, so we do not observe the true r in (2.14), but a precipitation observation r_o:

$$r_o = f(\mathbf{x}) + e_m + e_o, \tag{2.15}$$

where e_m and e_o are the model and observational error, respectively.

As we mentioned, a typical assimilation scheme would attempt to find $\hat{\mathbf{x}}$ to minimize (2.3), without explicit account of the distributions of e_m and e_o. It may even ignore the possibility of model error. But if f is nonlinear, the cost function we would minimize could have multiple minima. To investigate this and other issues we perform the following Bayesian analysis:

1. Assume the prior distribution for $\mathbf{x}$, written as $\pi(\mathbf{x})$, is Gaussian:

$$\mathbf{x} \sim \mathrm{N}\left[\begin{pmatrix} \mathbf{T}_p \\ \mathbf{q}_p \end{pmatrix}, \begin{pmatrix} \Sigma_T & 0 \\ 0 & \Sigma_q \end{pmatrix}\right].$$

2. Assume that the errors follow Gaussian distributions and are uncorrelated:

$$e_o \sim \mathrm{N}(0, \sigma_o^2), \qquad e_m \sim \mathrm{N}(0, \sigma_m^2), \qquad \mathrm{Cov}(e_o, e_m) = 0.$$

3. Attempt to assess the estimators based on the posterior $\pi(\mathbf{x}|r_0)$, which is derived from Bayes' theorem:

$$\pi(\mathbf{x}|r_o) \equiv \pi(\mathbf{T}, \mathbf{q}|r_o) = \frac{\pi(\mathbf{T}, \mathbf{q})f(r_o|\mathbf{T}, \mathbf{q})}{\int \pi(\mathbf{T}, \mathbf{q})f(r_o|\mathbf{T}, \mathbf{q}) \, d\mathbf{T} \, d\mathbf{q}}, \tag{2.16}$$

where $f(r_o|\mathbf{T}, \mathbf{q}) \sim \mathrm{N}(f(\mathbf{T}, \mathbf{q}), \sigma_o^2 + \sigma_m^2)$.

Although the Gaussian assumption in Step 1 may not be reasonable, along with $\mathbf{T}$ and $\mathbf{q}$ being uncorrelated, these assumptions reduce the complexity of the example and focuses the analysis on the nonlinearity in (2.15).

Although Step 2 violates our previous discussion of probable non-Gaussian error structure with precipitation data, we do this for simplicity, and to investigate what may be happening with presently used methods.

In Step 3 we have written down the calculation one would have to do to find the posterior distribution explicitly. However, there is no closed form for $\pi(\mathbf{x}|r_0)$ because the function $f(r_o|\mathbf{T}, \mathbf{q})$ is nonlinear. Therefore we approximate it using a Metropolis–Hastings algorithm [Car96, p. 173], which generates an approximate random sample from (2.16).

As we can see from the formulation of this problem, most of the effort goes into formulating all the distributions we must use. This is certainly not a trivial matter, and we have indicated why we made the simple choices of Gaussian distributions in our formulation. The specification of the prior model, and observational error should take into account many factors, such as the dynamics in question and properties of instruments used to collect the observations. However, once we have specified the distributions, the Bayes calculation is straightforward.

Once we have generated our approximate sample from (2.16), we make some qualitative assessment of the results. Specifically, we examine the approximate marginal posteriors (univariate and two-dimensional cases) for evidence of skewness or multimodality. The approximate posteriors were constructed using standard kernel density methods [Sim96, Chap. 3, 4]. We have done this under a number of different scenarios, and we present one of them here.

The example we present makes the following assumptions. We use an eight-level model with $r_o \approx 0.0023$, which is a measure of the mass of precipitation generated per mass of moist air within that volume. It can be related to actual rainfall amount in mm/hr, as explained by [Err99]. We assume that $\sigma_o^2 = \sigma_m^2 = 7 \times 10^{-6}$, while the choices for $\mathbf{T}_p$, $\mathbf{q}_p$, and the prior covariances are based on values used by [Err99]. The Metropolis–Hastings algorithm was run from its initial value for 10,000 iterations, and this was repeated for five different starting points. The final 50% of the samples generated from each run of the algorithm were retained and combined to give our approximate sample from (2.16).

Figure 1 plots the approximate marginal temperature posteriors versus the prior distributions used. In Figure 1, T1 corresponds to temperature at the level furthest from the surface (so furthest from the precipitation observation) while T8 is the lowest level. The plots indicate that there has been a shift in the modes of the posteriors from the priors, with the posterior mode being smaller at higher levels, but larger at lower levels. Some of these shifts have practical relevance, being on the order of 1°, and could have a noticeable effect in a large-scale numerical model. The posterior variances also change most at the

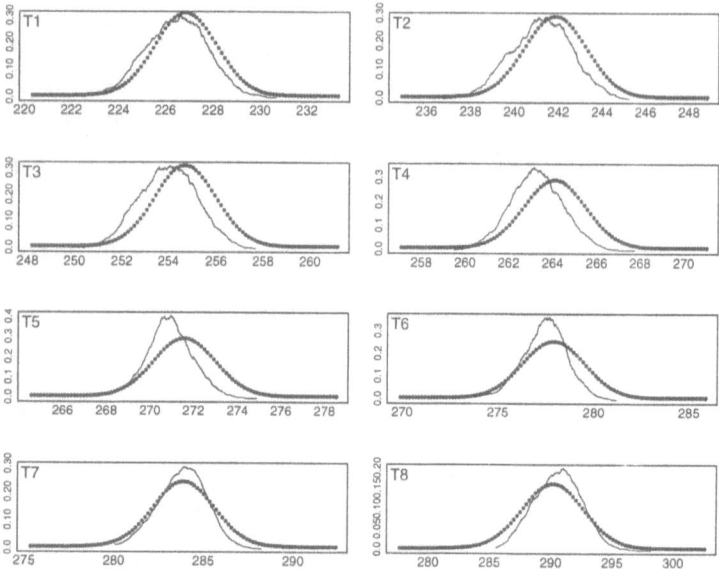

FIGURE 1. Prior and posterior distributions of temperatures, eight-level model. The solid line is the posterior density, and the dashed line is the prior.

lower levels. This is expected because the information from the observation is at the surface. Figure 2 shows the marginal posterior distribution of the specific humidity variables. We see that there is little shift from the priors used at the upper levels, with some increasing shift at the lower levels.

Figures 1 and 2 suggest that many of the posterior distributions appear Gaussian. To assess this in another way, we construct normal QQ-plots for the marginal posteriors, and present four of them in Figure 3. The two posterior distributions from the temperature variables do not display much non-Gaussian behavior. For the specific humidity variable, it does appear that the distribution in the lowest level is skewed.

Since an assimilation method often attempts to minimize a function in a multidimensional space, we constructed bivariate posterior distributions for the state variables. The presence of multimodality in these distributions may suggest that a minimization scheme could end up choosing a local, rather than global, minimum. Figure 4 contains a sample of four bivariate posteriors for the two variables, where we examine variables at neighboring levels in the atmosphere. We see that three of the plots exhibit multimodality, and this occurs with both the temperature and specific humidity variables.

To assess the significance of the multiple modes that appear to be present, we performed the following study for the upper-left plot in Figure 4. We took the estimated mean and covariance matrix of this bivariate distribution, and simulated 300 samples of size 500 from a Gaussian distribution with the same

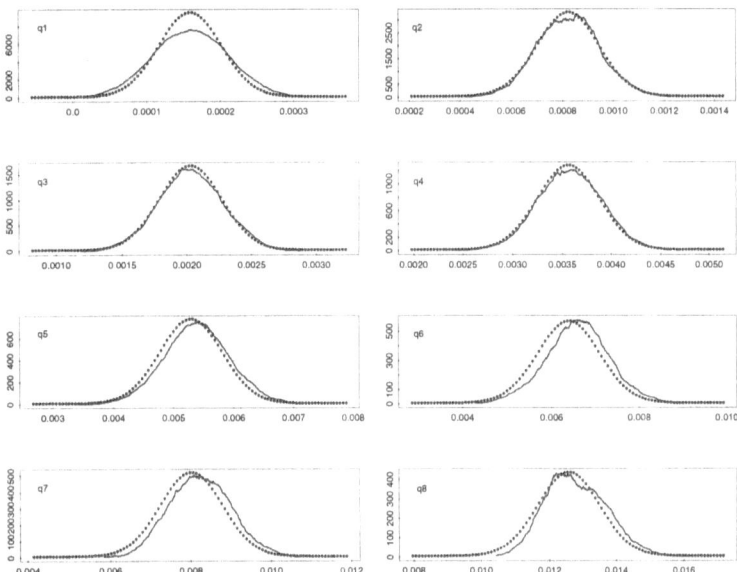

FIGURE 2. Prior and posterior distributions of specific humidity, eight-level model. The solid line is the posterior density, and the dashed line is the prior.

mean and variance. We then constructed kernel density estimators for each sample, and found how many of these exhibited multimodality. Fewer than 3% displayed multimodality to the extent we see in the plot, and so we consider this evidence that the bimodality we observe in this plot in Figure 4 is not an artifact of sampling.

In summary, by examining the marginal and posterior distributions of the state variables, we have a better understanding of the sensitivity of the penalized least squares approach to nonlinearities in the underlying model. We see that, in the majority of the posterior results, using a standard assimilation approach yielded posteriors that were approximately Gaussian. However, there is evidence to suggest that there is the potential for problems, particularly in the bivariate distributions.

There are natural extensions to this study. One is to allow for a more realistic error structure. Errico et al. [Err99] consider some non-Gaussian error structures, and do find bimodality present in some of their results. Another extension is to use an RAS scheme with more levels, in effect allowing a finer discretization of the atmosphere.

6 Conclusions

We have demonstrated that data assimilation has a natural correspondence with Bayesian methods, as it incorporates observations, models, and prior

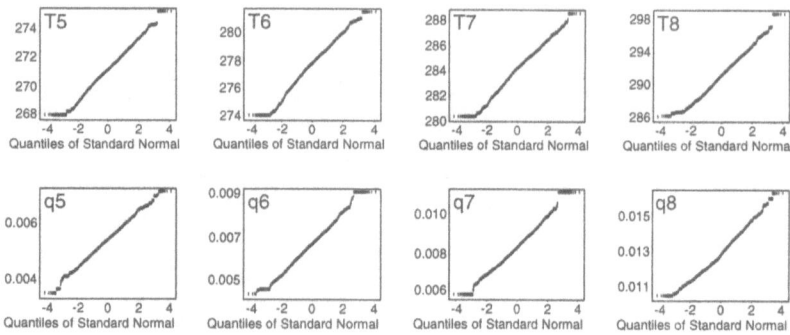

FIGURE 3. QQ-plots of selected temperature and specific humidity distributions, eight-level model.

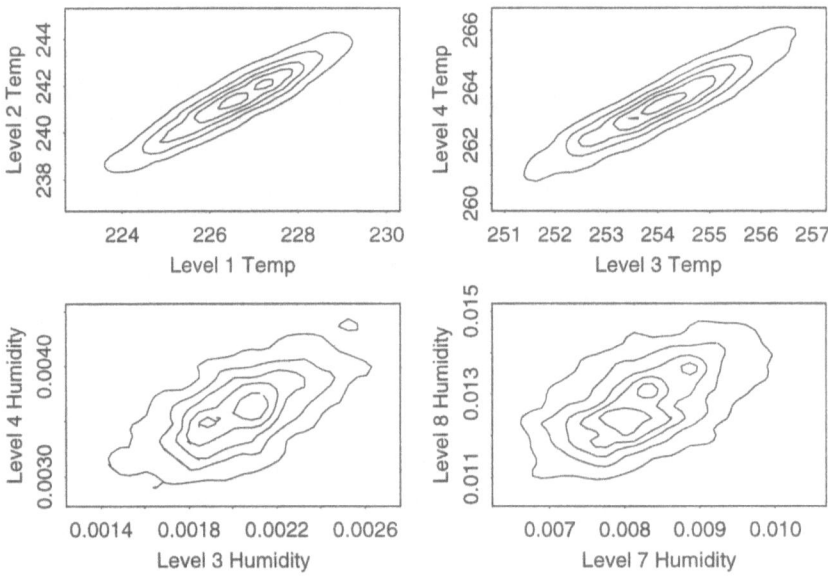

FIGURE 4. A sample of bivariate posterior distributions, eight-level model.

information in some optimal fashion. We have also shown that assimilation methods like ensemble forecasting are Monte Carlo methods that use a random sample to approximate a posterior distribution.

We have discussed some of the shortcomings that can be present in assimilation procedures, and our precipitation example showed the problems that can arise with a least squares procedure if there is only a nonlinear problem in the model. Therefore it seems natural that the knowledge of both disciplines can be combined to add to the understanding and development of assimilation methods. This could include correct ways of incorporating realistic error structure and studying methods of making confidence statements about the estimators that are produced by an assimilation procedure. The opportunities are present to produce creative, fundamental research in statistics and the geosciences through the study of data assimilation.

Acknowledgments

The author thanks Doug Nychka for his help in learning about the topic of data assimilation, and Ron Errico for his time and assistance on the precipitation problem. The author also thanks Mark Berliner, Tom Hamill, and Chris Snyder for many helpful discussions on ensemble forecasting.

Multivariate Spatial Models

J. Andrew Royle
U.S. Fish and Wildlife Service, Laurel, MD 20708, USA

1 Introduction

In studies involving spatial data, it is seldom the case that data for only a single process are collected. Typically, there is great expense associated with establishing spatial monitoring networks or other mechanisms of spatial data collection (e.g., satellites) and so measurements are usually made on two or more variables. Even when networks are established for purposes of monitoring a single physical process, there often exist alternative and disjoint networks that were established for monitoring related processes. Finally, it is seldom the case that a single physical process is uninfluenced by other processes, and so it would be inefficient to neglect information from these other processes even when one's objective centers on modeling or prediction of the single process. Thus, statistical techniques for multivariate spatial data are critical for effective modeling of spatial processes. The primary objectives of this chapter are to review existing multivariate spatial modeling strategies, discuss their strengths and weaknesses, and to discuss several atmospheric science problems that require multivariate spatial models. For a more thorough exposure to traditional geostatistical approaches, the reader is referred to Wackernagel [Wac95] and the review by Gotway and Hartford [Got96].

I will begin with a brief motivating example involving space–time modeling of ozone and meteorology followed by a review of univariate kriging. In Section 2, I discuss the multivariate extension known as cokriging. In Section 3, I will review the procedure known as *kriging with external drift* (KED) which is essentially the reduction of a multivariate problem to a simpler univariate problem. In Section 4, I present a hierarchical modeling approach that is not only quite general, but also avoids some of the difficulties and ambiguities inherent in the cokriging and KED approaches. For simplicity, these approaches are illustrated in the bivariate context. In Section 5, I discuss the general k-variate extensions, and address a few loose ends having to do with alternative modeling strategies. Finally, I will introduce two multivariate atmospheric science problems involving sea-surface wind fields and monthly minimum and maximum temperatures. These are not thoroughly analyzed here, but rather briefly introduced in order to give the reader some perspective in relation to the preceding material.

1.1 Motivating Example: Ozone and Meteorology

The primary message that I hope to leave the reader with is that one can simplify multivariate spatial modeling problems by the use of hierarchical models involving specification of one or more conditional models. To illustrate this general idea, consider the problem of building a multivariate space–time model for tropospheric ozone and meteorology. The analysis of tropospheric ozone data from EPA and state monitoring networks has become something of a cottage industry among environmental statisticians. While a lot of good work has been done on the subject, development of space–time models that incorporate covariates with stochastic structure has not been investigated. This problem is ideally suited for a conditional modeling framework because simplifying conditional assumptions can be made at several stages of the model. Namely, the "cause–effect" relationship between ozone and meteorology, and the dynamic aspect of "weather," can be combined to form a simple hierarchical space–time model for ozone. Moreover, this problem emphasizes an interesting potential application of multivariate models in general. That is, by accommodating the relationships among several processes, one can potentially use data on related processes that are less expensive to monitor, to inform about the primary variable. In the present context of ozone monitoring, it may be that collateral data on meteorological covariates are more efficient to collect on a cost/information basis, and the relationship between meteorology and ozone can be used to inform one about ozone levels at unmonitored sites.

A sketch of a possible multivariate hierarchical space–time model for ozone is presented here. Assume that the objective is modeling daily ozone and maximum temperature. Let $\mathbf{Y}_t$ be the vector of daily ozone observations at time t and let $\mathbf{Z}_t$ be the vector of maximum temperature observations. A reasonable assumption in this problem is that, conditional on present and past values of $\mathbf{Z}$, $\mathbf{Y}_t$ is independent of past values of $\mathbf{Y}$. That is, all of the temporal dynamics in ozone is propagated through the meteorological covariate. Thus the first stage in a space–time hierarchical model of ozone and temperature is

$$[\mathbf{Y}_t|\mathbf{Y}_{t-1},\ldots,\mathbf{Z}_t,\mathbf{Z}_{t-1},\ldots] = [\mathbf{Y}_t|\mathbf{Z}_t,\mathbf{Z}_{t-1},\ldots]. \tag{3.1}$$

The bracket notation $[\cdot]$ means "distribution of $\cdot$", and here the conditional distribution of $\mathbf{Y}_t$ is assumed to be a Gaussian distribution. One might further assume that $\mathbf{Y}_t$ is only dependent on the present value of $\mathbf{Z}$, or perhaps the present and $t-1$ values, in a linear fashion:

$$[\mathbf{Y}_t|\mathbf{Z}_t,\mathbf{Z}_{t-1},\ldots] = \mathrm{Gau}(\mathbf{A}_0\mathbf{Z}_t + \mathbf{A}_1\mathbf{Z}_{t-1}, \boldsymbol{\Sigma}_{y|z}). \tag{3.2}$$

The second stage of the model specifies a spatio–temporal model on the meteorological covariate. Supposing that a first-order Markovian assumption is valid for $\mathbf{Z}_t$, then this stage of the model is completely specified by the pieces

$$[\mathbf{Z}_t|\mathbf{Z}_{t-1}] = \mathrm{Gau}(\mathbf{B}_1\mathbf{Z}_{t-1}, \boldsymbol{\Sigma}_z) \qquad \text{for} \quad t = 1,...,T. \tag{3.3}$$

Spatial structure can be parameterized at both stages of this model: conditional on temperature, there exists "residual" spatial correlation in ozone ($\Sigma_{y|z}$ of (3.2)) and, similarly, temperature is spatially correlated conditional on its past values (Σ_z of (3.3)).

Thus, the hierarchical paradigm greatly simplifies the multivariate space–time nature of this problem. In essence, a univariate *space–time* model is constructed for temperature and this drives a univariate *spatial* ozone model. This brief sketch leaves out much detail. In general, the relationships need to be made precise, by specification of parameter matrices $\mathbf{A}_0, \mathbf{A}_1, \ldots, \mathbf{B}_1$ and a covariance structure implicit in $\Sigma_{y|z}$ and Σ_z. These details are beyond the scope of this chapter.

1.2   Optimal Spatial Prediction—Kriging

Spatial prediction in the atmospheric sciences dates to the 1960s [Gan65] and is usually known as *optimal interpolation*. (See the recent translation of Kagan [Kag97] for many references on the early developments in atmospheric science.) Spatial prediction is referred to as *kriging* in the geostatistical literature due to early work in mining engineering by Krige [Kri51]. Statisticians know this as *best linear unbiased prediction* (BLUP), and the statistical formulation predates those of either field. The interested reader is referred to [Cre90] for an historical treatment of the subject.

Kriging is generally presented as a multistep procedure; estimating covariances or other second-moment quantities, selecting and fitting a theoretical model to those estimates, and computing predictions based on a fitted covariance model. Since the details of estimation and prediction are presented in depth elsewhere (e.g., [Cre93]), I will focus on the kriging *model* and its extension to the multivariate setting.

1.3   Universal Kriging

Universal kriging [Cre93, p. 120], specifies the model through linear means and a covariance function. Although predictions made under this model are optimal under a Gaussian assumption with known covariances, such precise distributional assumptions are seldom made because there is a long history of anxiety in the geostatistical community over the Gaussian assumption. In keeping with this convention, random variables in this chapter will be characterized by their first and second moments alone, because this is sufficient to develop the BLUP. Thus, I might casually refer to the "distribution" of a random variable, in which case I mean its first and second moments.

The universal kriging model assumes that the process $Y(s)$ has an expected value $E[Y(s)] = \mu(s)$, and covariance function

$$\text{Cov}(Y(s), Y(s')) = \sigma^2 k_y(s, s'),$$

where σ^2 is the process variance, k_y is the correlation function of the Y process, and s and s' are any two locations in the region over which $Y(s)$ is defined. It is assumed that $\mu(s)$ is a linear function:

$$E[Y(s)] = \sum_{j=1}^{p} \beta_j x_j(s),$$

where $x_j(s)$ are fixed regression functions. Typically, stationarity (k_y is only a function of $s - s'$) and isotropy (k_y is only a function of $||s - s'||$) constraints are imposed on k_y for practical reasons and so k_y is often parameterized by a low-dimensional vector of parameters, say η. Strict isotropy is easily relaxed in favor of various anisotropic parameterizations such as elliptic anisotropy (see [Zim93] for a discussion of anisotropy in geostatistics). Nonstationary parameterizations of the correlation function are scarce in the literature and so stationarity is often seen as a required assumption, at least locally. The idea that local stationarity can be exploited to improve prediction of globally nonstationary processes is central to the work of Haas [Haa90], [Haa95] in his "moving-window" kriging procedure. In addition to parameters describing the shape of this correlation function, η might contain a parameter to account for measurement error (or "nugget" in geostatistical parlance) which amounts to a discontinuity in the covariogram at lag distances of 0.

The universal kriging problem is a prediction of Y at an unmonitored location, say $Y(s_o) \equiv Y_o$ given a vector of n observations, $\mathbf{Y}_d$. Let $\mathbf{x}'_o$ be the vector $(x_1(s_o), \ldots, x_p(s_o))$, and let $E[\mathbf{Y}_d] = \mathbf{X}\beta$. Given the correlation function k_y, one can compute the quantities $\text{Cov}(Y_o, \mathbf{Y}_d) = \sigma^2 \mathbf{k}'_y$ and $\text{Cov}(\mathbf{Y}_d, \mathbf{Y}_d) = \sigma^2 \mathbf{K}_y$ (note that I am using k_y to represent a function, whereas boldface lower- and uppercase letters represent vectors and matrices, respectively). Then, using the generalized least squares estimator of β, $\hat{\beta} = (\mathbf{X}'\mathbf{K}_y^{-1}\mathbf{X})^{-1}\mathbf{X}'\mathbf{K}_y^{-1}\mathbf{Y}_d$, the predictor of Y_o is

$$\hat{Y}_o = \mathbf{x}'_o\hat{\beta} + \mathbf{k}'_y\mathbf{K}_y^{-1}(\mathbf{Y}_d - \mathbf{X}\hat{\beta}). \tag{3.4}$$

This prediction can be derived by minimizing the mean-squared error of $\hat{Y}_o$ subject to the estimator being unbiased and a linear function of $\mathbf{Y}_d$. Alternatively, $\hat{Y}_o$ can be interpreted as the conditional expectation of Y_o given $\mathbf{Y}_d$ under a Gaussian assumption on the vector $\mathbf{Y} = (Y_o, \mathbf{Y}_d)$. The prediction variance, or minimized mean-squared error (MSE) of the predictor is (e.g., [Chr91, p. 269]):

$$\text{Var}(\hat{Y}_o - Y_o) = \sigma^2 - 2\gamma'\mathbf{k}_y + \gamma'\mathbf{K}_y\gamma, \tag{3.5}$$

where γ is given by the expression $\gamma = \mathbf{x}'_o(\mathbf{X}\mathbf{K}_y^{-1}\mathbf{X})^{-1}\mathbf{X}'\mathbf{K}_y^{-1} + \mathbf{k}'_y\mathbf{K}_y^{-1}(\mathbf{I} - \mathbf{X}(\mathbf{X}'\mathbf{K}_y^{-1}\mathbf{X})^{-1}\mathbf{X}'\mathbf{K}_y^{-1})$.

The kriging predictor is the "optimal" predictor or BLUP only for the case in which the second-moment structure of the process is known. In practice, second-moment parameters (i.e., the covariance function) are always estimated and thus, strictly speaking, the kriging predictor is not optimal. See [Zim92] and [Chr91, p. 276] for a discussion of the impact of using estimated second-moment parameters on prediction.

The preceding discussion has focused on prediction using a covariance function (or *covariogram*) to describe the second-moment structure. An equivalent development exists using the variogram, which is defined as $\gamma(s, s') = E[(Y(s) - Y(s'))^2]$ (see [Cre93, p. 123] and [Chr91, p. 270]). Practically speaking, there are few differences between using the covariogram or variogram, although this seems to be a matter of some debate in the geostatistical community. However, many statisticians and atmospheric scientists prefer to quantify second-moment dependence using the covariogram, whereas traditional geostatisticians tend to prefer use of the variogram.

1.4 Multivariate Approaches

Let $Y_1(s), Y_2(s), \ldots, Y_p(s)$ be a set of p variables and assume that predictions of one or more of the variables are desired. For clarity, it is convenient to examine the bivariate case $(p = 2)$. In this case, the two processes are indicated with different letters, say $Y(s)$ and $Z(s)$. The traditional modeling approaches for bivariate (and more general multivariate) situations are straightforward extensions of univariate kriging. The *cokriging* model (Section 2) is formulated through direct and complete specification of the joint second moment structure of the processes, including *cross-covariance* structure. The *kriging with external drift* model (Section 3) reduces the amount of second-moment modeling required, but at the expense of producing a model capable of joint prediction. A third approach based on hierarchical modeling (Section 4) takes the middle road between these two approaches by *indirectly* modeling cross-covariance structure in a conditional mean model, yet retaining joint prediction capabilities.

2 Cokriging

The cokriging model develops the best linear unbiased predictor from direct specification of the joint distribution (or the first and second moments) of the multivariate process. For example, in the bivariate case, we assume linear means:

$$
\begin{aligned}
E[Y(s)] &= \mathbf{x}_y'(s)\boldsymbol{\beta}_y, \\
E[Z(s)] &= \mathbf{x}_z'(s)\boldsymbol{\beta}_z,
\end{aligned}
$$

where $\mathbf{x}_y$ and $\boldsymbol{\beta}_y$ are p_y vectors and $\mathbf{x}_z$ and $\boldsymbol{\beta}_z$ are p_z vectors. The covariance structure is defined by a set of functions:

$$\begin{aligned}
\mathrm{Cov}[Y(s), Y(s')] &= \sigma_y^2 k_y(s, s'), \\
\mathrm{Cov}[Z(s), Z(s')] &= \sigma_z^2 k_z(s, s'), \\
\mathrm{Cov}[Y(s), Z(s')] &= \sigma_{yz}^2 k_{yz}(s, s'),
\end{aligned}$$

where the last quantity is referred to as the *cross-covariance* function. The objective is the prediction of $Y_o \equiv Y(s_o)$ and $Z_o \equiv Z(s_o)$ from data $\mathbf{Y}_d$ and $\mathbf{Z}_d$. For simplicity, assume that both variables are observed at n locations, and so these vectors are $n \times 1$.

To specify the cokriging predictor, define the partitioned matrices:

$$\begin{aligned}
\Sigma_1 &= \left(\begin{array}{cc} \mathrm{Cov}(Y_o, \mathbf{Y}_d)' & \mathrm{Cov}(Y_o, \mathbf{Z}_d)' \\ Cov(Z_o, \mathbf{Y}_d)' & Cov(Z_o, \mathbf{Z}_d)' \end{array} \right)_{2 \times 2n} \\
&= \left(\begin{array}{cc} \mathbf{k}_y' & \mathbf{k}_{yz}' \\ \mathbf{k}_{zy}' & \mathbf{k}_z' \end{array} \right)
\end{aligned}$$

and

$$\begin{aligned}
\Sigma_2 &= \left(\begin{array}{cc} \mathrm{Cov}(\mathbf{Y}_d, \mathbf{Y}_d) & \mathrm{Cov}(\mathbf{Y}_d, \mathbf{Z}_d) \\ \mathrm{Cov}(\mathbf{Z}_d, \mathbf{Y}_d) & \mathrm{Cov}(\mathbf{Z}_d, \mathbf{Z}_d) \end{array} \right)_{2n \times 2n} \\
&= \left(\begin{array}{cc} \mathbf{K}_y & \mathbf{K}_{yz} \\ \mathbf{K}_{yz}' & \mathbf{K}_z \end{array} \right).
\end{aligned}$$

Under this model, the cokriging predictor (or BLUP, equivalent to the conditional mean of (Y_o, Z_o) given $(\mathbf{Y}_d, \mathbf{Z}_d)$ under a Gaussian assumption) is then

$$\left(\begin{array}{c} \hat{Y}_o \\ \hat{Z}_o \end{array} \right) = \left(\begin{array}{c} \mathbf{x}_y'(s_o)\beta_y \\ \mathbf{x}_z'(s_o)\beta_z \end{array} \right) + \Sigma_1 \Sigma_2^{-1} \left(\begin{array}{c} \mathbf{Y}_d - \mathbf{X}_y\beta_y \\ \mathbf{Z}_d - \mathbf{X}_z\beta_z \end{array} \right)$$

which is identical to the equation given by (3.4) for the univariate case. The expression for the prediction variance *matrix* is again very similar to that for the universal kriging case. For details and general treatments of cokriging see the detailed papers by [Mye82] and [Ver93].

In practice, the critical task in cokriging is estimation of the covariance functions for Y and Z, and the cross-covariance function. Of course, the joint variance–covariance matrix for any vector of Y's and Z's must be positive definite, and imposing this constraint on arbitrary functions and their estimates is not a simple matter (see [Mye91] and [Ver98]). Indeed, in many applications the functions appear to be estimated more or less independently of one another, which does not guarantee that the resulting estimates are valid. Occasionally, a posteriori "checking" that resulting estimates yield positive definite covariance matrices is done, which induces some arbitrariness into the whole modeling exercise. Ver Hoef and Barry [Ver98] suggest a general approach for nonparametric estimation of the joint covariance structure that does lead to valid estimates.

Perhaps the major limitation of cokriging is that known dependencies between the *means* of the processes are not easy to accommodate. Cokriging seeks to model relationships between processes in the second-moments, and it is likely that most applied scientists think about relationships between variables in terms of first-moment relationships (e.g., regression functions). When this is the case, modeling relationships in terms of cross-covariance functions makes the modeling task more difficult.

3   Kriging with External Drift

Another popular approach for analyzing multivariate data is known as "kriging with external drift" (KED) in the geostatistical literature [Ahm87], [Got96]. This procedure is not, strictly speaking, multivariate. Rather, it is used in situations in which one is primarily interested in a single variable, say Y, but has observations of one or more covariates (say Z in the bivariate case), to aid in making predictions of Y. The key idea is a reduction of the multivariate model to a simpler univariate conditional (regression) model, specifying the conditional dependence of Y on Z. KED is discussed here for completeness, but also because some of the difficulties encountered in its application motivate the need for more general conditionally specified models (Section 4).

For the bivariate case, consider the regression model:

$$Y(s)|Z(s) = \beta_0 + \beta_1 Z(s) + e(s),$$

where $e(s)$ is a correlated spatial process. This is a simple universal kriging model except that the spatially varying mean is a function of the random variable $Z(s)$ rather than deterministic functions of spatial coordinates (hence the terminology "external drift"). One appealing aspect of the KED model is that it does allow one to directly parameterize known conditional relationships between the variables. For example, chemical reactions leading to the formation of tropospheric ozone are moderated to a large extent by meteorological conditions, thus providing the basis for a conditional model specification.

Application of KED is limited when the goal is prediction of both Y and Z. Also, KED does not directly accommodate stochastic structure in Z. As a consequence, there is some difficulty in applying KED in the presence of missing values of the covariate. This is typically stated as requirements on the sampling of the covariate process: $Z(s)$ must be *exhaustively* sampled so that values of Z are available wherever Y predictions are desired, and Y and Z must be observed at the same points in space (i.e., they are paired, coincident, or colocated) in order to estimate parameters. These are serious limitations. For example, a monitoring network might be established to measure an air pollutant (e.g., ozone), and meteorological covariate information might be available from a second network of stations established without regard to the monitoring of air pollution. One would like a model that can handle data collected under this situation, while at the same time being able to accommodate known conditional relationships. An ad hoc solution to dealing with problems that deviate from these sampling requirements is to estimate the missing Z values by interpolation from available data (see [Got96], [Ahm87], [Haa96]). Failure to account for interpolation error in such procedures will have the impact of biasing prediction variances. Since the historical rationale for kriging over other techniques (e.g., spline smoothing) is that measures of prediction uncertainty are more easily produced, this is troublesome.

3.1. Prediction Variance Misspecification in the KED Model

Bias in the KED prediction variance when predicted values of the covariate are used is due to ignoring the variation in imputation of the missing covariate. The model specifies a conditional model relating $Y(s_o)$ to $Z(s_o)$, and thus $Z(s_o)$ appears explicitly in the predictor, and the prediction variance is reduced accordingly. When this value is missing, the *true* optimal predictor should be produced by conditioning only on the available *data*. In other words, using a predictor obtained by substituting filled-in values of Z into the KED predictor leads to a predictor that is not optimal. More importantly, the KED prediction variance computed assuming that the filled-in missing $Z(s_o)$ is the true value is not the correct variance.

To clarify, let $P(Y(s_o)|Z(s_o), \mathbf{Y}_d, \mathbf{Z}_d)$ denote the KED predictor and let the corresponding KED prediction variance be denoted by $V(Y(s_o)|Z(s_o), \mathbf{Y}_d, \mathbf{Z}_d)$. Let $P_{\text{plug}}(Y(s_o)|\tilde{Z}(s_o), \mathbf{Y}_d, \mathbf{Z}_d)$ denote a "plug-in" predictor, where $\tilde{Z}(s_o)$ is the filled-in value of $Z(s_o)$. Let $V(P_{\text{plug}}(Y(s_o)))$ denote the variance of this predictor, computed assuming $\tilde{Z}(s_o)$ to be a random quantity. Assume that $Z(s_o)$ is filled-in using an arbitrary predictor, but one that has variance $\tau^2 > 0$. The problem arises because $V(Y(s_o)|Z(s_o), \mathbf{Y}_d, \mathbf{Z}_d)$ is often used *incorrectly* as the predictive variance associated with $P_{\text{plug}}(Y(s_o)|\tilde{Z}(s_o), \mathbf{Y}_d, \mathbf{Z}_d)$.

Now, the true *optimal* predictor and its associated predictive variance when $Z(s_o)$ is missing can be obtained by integrating the unobserved quantity $Z(s_o)$ from the problem using rules of iterated conditional variance [Cas90, Chap. 4]. This variance is

$$V(Y(s_o)|\mathbf{Y}_d, \mathbf{Z}_d) = \beta_1^2 \operatorname{Var}[Z(s_o)|\mathbf{Y}_d, \mathbf{Z}_d] + V(Y(s_o)|Z(s_o), \mathbf{Y}_d, \mathbf{Z}_d). \quad (3.6)$$

Thus, we see that the optimal predictive variance is larger than the misspecified KED prediction variance by addition of the quantity $\beta_1^2 \operatorname{Var}[Z(s_o)|\mathbf{Y}_d, \mathbf{Z}_d]$. Unfortunately, the quantity $\operatorname{Var}[Z(s_o)|\mathbf{Y}_d, \mathbf{Z}_d]$ cannot be computed under the KED model since it does not admit any stochastic structure for Z.

Also note that the KED variance—i.e., $V(Y(s_o)|Z(s_o), \mathbf{Y}_d, \mathbf{Z}_d)$—is not the variance of the misspecified KED predictor. Under mild assumptions, it can be shown that the true variance of the misspecified KED predictor is

$$V(P_{\text{plug}}(Y(s_o))) = \beta_1^2 \tau^2 + V(Y(s_o)|Z(s_o), \mathbf{Y}_d, \mathbf{Z}_d). \tag{3.7}$$

Again, we observe that this quantity is larger than the naively stated (i.e., misspecified) KED prediction variance. Additionally, this quantity must be greater than the variance of the optimal predictor given by (3.6).

These observations can be summarized as follows. The misspecified prediction variance **underestimates** the true prediction variance by a term proportional to β_1^2. So, the prediction variance bias increases with the strength of the linear relationship (holding marginal variances fixed) between Y and Z. Since this is precisely when one might wish to use a KED-like model for prediction, accounting for missing values of the covariate is crucial to properly assess the variability of predictions.

Strictly speaking, assumptions on the sampling of Z are not necessary and in fact are due to specification of an incomplete model for these more general situations. A simple modification to the model, by incorporating stochastic structure on the Z process, while maintaining the conditional model specification, allows for arbitrary sampling conditions. This is the essence of the hierarchical model presented in the next section.

4   A Hierarchical Model

Royle and Berliner [Roy99b] recently proposed a simple hierarchical model for multivariate spatial processes. This model shifts away from the cokriging paradigm of modeling relationships in the second moments to modeling those relationships in the first moments. As a consequence, specification and estimation of cross-covariance models is unnecessary. However, the model is similar to cokriging in that a full joint model is specified for the process, enabling joint prediction and inference, but it also allows one to accommodate conditional relationships between processes as in KED. For this discussion, the notation $\mathbf{Z} \sim (\boldsymbol{\mu}, \boldsymbol{\Sigma})$ will mean that $\mathbf{Z}$ has mean $\boldsymbol{\mu}$ and variance–covariance matrix $\boldsymbol{\Sigma}$.

In the bivariate case, let $\mathbf{Y} = (Y(s_o), \mathbf{Y}_d)$ and $\mathbf{Z} = (Z(s_o), \mathbf{Z}_d)$ (both $(n + 1) \times 1$ vectors, although all following arguments generalize to $\mathbf{Y}$ and $\mathbf{Z}$ of differing length). The first stage assumes a conditional model for $\mathbf{Y}$ given $\mathbf{Z}$:

$$\mathbf{Y}|\mathbf{Z} \sim (\mathbf{X}_y \boldsymbol{\beta}_y + \mathbf{BZ}, \boldsymbol{\Sigma}_{y|z}), \tag{3.8}$$

where $\mathbf{X}_y$ is a fixed $(n + 1) \times p_y$ matrix and $\boldsymbol{\beta}_y$ is a p_y vector of parameters. The elements of $\mathbf{X}_y$ are typically chosen to model "spatial drift." The matrix $\boldsymbol{\Sigma}_{y|z}$ is an $(n+1) \times (n+1)$ conditional covariance matrix, which is assumed to depend on a low-dimensional parameter vector, say $\boldsymbol{\eta}_{y|z}$. To avoid confusion, however, this dependence is suppressed. The key feature is that $\boldsymbol{\Sigma}_{y|z}$ is typically easier to model than is the marginal covariance $\boldsymbol{\Sigma}_y$, because some spatial dependence is explained by conditioning on the covariate. Finally, $\mathbf{B}$ is a matrix of parameters relating the Z process to the Y process. In general, one would consider parameterizations of $\mathbf{B}$ that involve a relatively low-dimensional vector of *regression* parameters $\boldsymbol{\theta}$, say $p \times 1$. For clarity, this dependence will be suppressed.

It is convenient to assume that $\mathbf{BZ}$ is *linear* in $\boldsymbol{\theta}$. This means that there exists an $(n+1) \times p$ *matrix* $\boldsymbol{\Psi}$ (whose elements only depend on $\mathbf{Z}$) such that

$$\mathbf{BZ} = \boldsymbol{\Psi\theta}. \tag{3.9}$$

This requirement enables the use of generalized least squares estimation. See [Roy99b] for further details and some examples of $\mathbf{B}$ matrices that are linear in $\boldsymbol{\theta}$ and the corresponding $\boldsymbol{\Psi}$. One such parameterization is given in Section 6.1 below.

The conditional mean model can be written as

$$E(\mathbf{Y}|\mathbf{Z}) = \mathbf{X}_y\boldsymbol{\beta}_y + \boldsymbol{\Psi\theta} \tag{3.10}$$

which looks similar to the KED model. Indeed, if $\mathbf{B} = \beta_1\mathbf{I}$ where $\mathbf{I}$ is the identity matrix, then $\boldsymbol{\Psi} = \mathbf{Z}$ and $\theta = \beta_1$ and this is precisely the conditional mean of the KED model.

The second stage of the model specifies a marginal spatial model for the covariate $\mathbf{Z}$:

$$\mathbf{Z} \sim (\mathbf{X}_z\boldsymbol{\beta}_z, \boldsymbol{\Sigma}_z), \tag{3.11}$$

where $\mathbf{X}_z$ is a fixed $(n+1) \times p_z$ matrix and $\boldsymbol{\beta}_z$ is a p_z vector of parameters. As above, the elements of $\mathbf{X}_z$ are typically chosen to model spatial drift. Here $\boldsymbol{\Sigma}_z$ is an $(n+1) \times (n+1)$ covariance matrix which is the *marginal* covariance matrix of $\mathbf{Z}$. Again, this matrix is parameterized by a low-dimensional vector $\boldsymbol{\eta}_z$. Note that the conditional (3.8) and marginal (3.11) do completely specify the joint distribution of $\mathbf{Y}$ and $\mathbf{Z}$.

It is straightforward to identify the *implied* joint means and covariances of $\mathbf{Y}$ and $\mathbf{Z}$ under the hierarchical model. In particular, it is instructive to examine the implied joint variance–covariance structure which is

$$\mathrm{Var}\begin{pmatrix} \mathbf{Y} \\ \mathbf{Z} \end{pmatrix} = \begin{pmatrix} \mathbf{K}_{y|z} + \mathbf{BK}_z\mathbf{B}' & \mathbf{BK}_z \\ (\mathbf{BK}_z)' & \mathbf{K}_z \end{pmatrix}. \tag{3.12}$$

Several interesting observations can be made based on this. First, the cross-covariance structure is simply a scaled version of the marginal covariance of $\mathbf{Z}$. Second, depending on the parameterization of $\mathbf{B}$, the implied cross-covariance and the marginal covariance of $\mathbf{Y}$ can be nonstationary. Finally, this joint covariance structure is guaranteed to be positive definite. See [Roy99b] for further details. The conditional model does lead to valid joint covariance structure but it comes at the expense of restricting its form to some extent. Although the conditional specification is sufficient for producing valid joint covariance structure, it is by no means necessary. While this restriction may be somewhat limiting in general, it is certainly realistic for situations in which conditional relationships exist between the variables. Estimation and prediction under the hierarchical model is relatively straightforward. Royle and Berliner [Roy99b] suggest an iterative approach that begins with imputation of missing values (if necessary), followed by generalized least squares estimation of parameters, and then estimation of covariance parameters from residuals.

In general, there may not be a natural ordering of this hierarchical model (i.e., Is Y conditional on Z? or Is Z conditional on Y?). In some cases, known physical dependencies will suggest which process should be modeled in the conditional model. For cases in which conditioning is not suggested by the nature of the problem, one might choose the ordering that produces the simplest covariance structure.

In summary, the hierarchical spatial model has several advantages over cokriging. First, the task of modeling cross-covariances is avoided. Second, relationships between variables are parameterized in the mean. This is natural for situations when known "cause–effect" relationships exist. Finally, the conditional formulation permits great generality in the types of processes that may be modeled, particularly if one adopts a Bayesian approach. For example, non-Gaussian models are more tractable in a conditional modeling framework. One might have a Poisson or Bernoulli Y process that is independent conditional on a Gaussian covariate Z. This conditional structure greatly facilitates estimation and prediction (e.g., via MCMC techniques). [Dig98] analyze such models for that case where Y is a univariate non-Gaussian spatial process, conditional on the *latent* Gaussian process Z. Thus, they essentially coerce a non-Gaussian univariate problem into a multivariate one in order to facilitate more realistic statistical modeling.

5   Miscellaneous Topics

5.1   K-Variate Extensions

Extensions of models presented in the previous sections to general k-variate problems is conceptually straightforward. However, the additional notation required for the general presentation obscures the simplicity behind these methods. In general, cokriging would require specification of $k(k + 1)/2$ cross-covariance functions which can be a cumbersome task. See [Mye82] and [Ver93] for the more general formulation of cokriging. The hierarchical approach of Section 4 is easier to apply in many situations, particularly when the scientific context can be exploited to suggest specific conditional models. Thus, effective implementation of the hierarchical approach usually requires subject-matter-based modeling decisions in order to develop the hierarchy since many possible formulations of the hierarchical model are possible. [Roy99b] elaborate on this and [Roy99c] provide an example of a trivariate Bayesian hierarchical spatial model involving wind fields and sea-surface pressure; described in Section 6.1.

5.2   Latent Process Models

Useful multivariate spatial models can be formulated by expressing the variables as functions of latent processes. One simple bivariate model for the processes $Y(s)$ and $Z(s)$ is

$$\begin{aligned}
Y(s) &= V(s) + W_y(s), \\
Z(s) &= V(s) + W_z(s),
\end{aligned} \tag{3.13}$$

where $V(s)$ is a random field with covariance function k_v common to both the Y and Z processes and W_y and W_z are independent random fields with covariance functions k_y and k_z. Under this model, the marginal covariance functions of Y and Z are sums of the covariance functions of the component latent processes. That is, $\mathrm{Cov}(Y(s), Y(s')) = k_v(s, s') + k_y(s, s')$ and $\mathrm{Cov}(Z(s), Z(s')) = k_v(s, s') + k_z(s, s')$. The cross-covariance function between Y and Z is simply $\mathrm{Cov}(Y(s), Z(s')) = k_v(s, s')$. See [Maj97] for an application of the model given in (3.14). The latent process model approach is appealing for general k-variate problems especially when physical constraints can be imposed on the observed field. For example, if the processes have common physical *forcings*, so that they have one or more sources of variation in common, then this can easily be expressed in terms of latent processes. They can also be useful for modeling non-Gaussian processes as demonstrated by [Dig98].

5.3   Modeling Orthogonal Contrasts

Much of the burden in constructing multivariate spatial models lies in formulating the dependencies among the different variables. To circumvent this, one might consider modeling functions of the variables that have a simplified dependence structure. That is, instead of modeling Y and Z directly, consider modeling linear combinations of Y and Z, say:

$$U(s) = a_1 Y(s) + a_2 Z(s)$$

and

$$V(s) = b_1 Y(s) + b_2 Z(s),$$

such that the "new" variables U and V are independent or have a simplified covariance structure.

One obvious choice of transforming the original process is orthogonal contrasts. That is, choose the vectors $\mathbf{a}' = (a_1, a_2)$ and $\mathbf{b}' = (b_1, b_2)$ such that $\mathbf{a}'\mathbf{b} = 0$. In cases where Y and Z have a shared component of spatial variation (such as under a latent process model), this can have the effect of yielding U and V processes with simplified cross-covariance structures. An approach of choosing a weighting scheme that is not necessarily orthogonal, but such that the transformed processes are uncorrelated (or nearly so) seems appealing from a modeling perspective since it greatly simplifies the task of modeling second-moment structures. While this idea has not been explored in any detail, the motivation arises from modeling wind and temperature fields (see Sections 6.1 and 6.2 below), for which it is more natural to model particular transformations of the data.

5.4   Multivariate Space–Time Models

Up to this point, we have ignored the temporal component inherent in all atmospheric and environmental processes. The simplest way to handle temporal replications of spatial data is to assume that they are temporally independent and that the spatial structure does not change through time. While this makes it possible to better estimate the spatial correlation structure, it is not generally efficient since it ignores the temporally dynamic aspect of these processes and neglects temporal correlation that can greatly aid in prediction.

Extension of kriging to space–time problems has been discussed by many authors (e.g., [Bil85], [Rou89], [Rou90a], [Rou90b]). A recent review can be found in [Kyr99]. These geostatistical approaches either treat time as another spatial dimension, or they treat the temporal observations as separate variables. Neither approach is entirely satisfactory. The former is awkward because it is unclear how to define "distance" in space–time, while analysis under the latter approach becomes computationally prohibitive when data are collected at many time points. Both approaches ignore the temporally dynamic nature of space–time processes. More generally, specification of realistic space–time covariance models is difficult. The standard models are *separable* in the sense that the space–time covariance function is either a product or sum of the spatial and temporal covariance functions (note that the latter can be derived from a latent process model specification). A recent innovation is [Cre99] who propose a flexible class of space–time covariance models. These show promise toward facilitating the kriging approach to space–time modeling.

In contrast to the difficulties of space–time kriging approaches, the hierarchical model of Section 4 extends readily to spatio–temporal processes. To motivate this extension, consider the problem involving univariate space–time data where the "variables" are indexed by time. Thus

$$\mathbf{Y}_t = (Y(s_1, t), Y(s_2, t), \ldots, Y(s_n, t)), \qquad t = 1, 2, \ldots, N,$$

are *replications* of the *same* type of variable, e.g., observations of an ozone, or temperature field over time. (This is in contrast to the usual cokriging situation where the observations are of different processes.) To analyze these univariate space–time datasets in a geostatistical paradigm, one might use cokriging, treating each of the $\mathbf{Y}_t, t = 1, 2, \ldots, N$, as distinct variables. However, as a consequence of these variables being observations of the *same* process, one might consider covariance models that are different from those for observations of *different* processes. For example, a separable model might make sense in the replicate situation because it is plausible that each spatial field has the same underlying covariance structure. This is not evident in the latter case, where the observations are not of the same process. Similarly, time replications of the same process enable one to consider temporal averages in order to estimate nonstationary spatial covariances . In the traditional cokriging setting involving observations of different processes, this would be absurd.

Of more importance here is that conditional modeling is very natural in space–time situations but not in the usual multivariate setting. Indeed, the formulation of conditional models is the rule in most applications that emphasize the temporal variation of a process. That is, for estimation and prediction, it is standard (e.g., the Kalman filter) to formulate the joint model of the space–time process, $[\mathbf{Y}_t, \mathbf{Y}_{t-1}, \ldots, \mathbf{Y}_1]$ via the simpler pieces

$$[\mathbf{Y}_t|\mathbf{Y}_{t-1}, \mathbf{Y}_{t-2}, \ldots, \mathbf{Y}_1] = [\mathbf{Y}_t|\mathbf{Y}_{t-1}], \qquad (3.14)$$

the equality being a consequence of a first-order Markov assumption.

In principle, this extends readily to multivariate settings. For example, to analyze

$$[\mathbf{Y}_t, \mathbf{Y}_{t-1}, \ldots, \mathbf{Y}_1, \mathbf{Z}_t, \mathbf{Z}_{t-1}, \ldots, \mathbf{Z}_1],$$

one could specify a model of Y *conditional on* Z:

$$[\mathbf{Y}_t, \mathbf{Y}_{t-1}, \ldots, \mathbf{Y}_1|\mathbf{Z}_t, \mathbf{Z}_{t-1}, \ldots, \mathbf{Z}_1], \qquad (3.15)$$

and a "marginal" space–time model for Z: $[\mathbf{Z}_t, \mathbf{Z}_{t-1}, \ldots, \mathbf{Z}_1]$. Both pieces could be simplified further by additional conditional assumptions (i.e., Markov assumptions as in the univariate space–time case) as was done in Section 1.1. Thus, the model sketch provided in Section 1.1 is just a synthesis of the multivariate hierarchical spatial model of Section 4 with the more traditional hierarchical (univariate) space–time modeling framework arising from (3.14).

In conclusion, a hierarchical formulation of space–time models is very natural and can often be used in conjunction with physical knowledge of the problem to produce simpler models. [Wik99a] (this volume) discusses the hierarchical perspective on space–time modeling in more detail.

6   Applications

6.1   Modeling Wind Fields

Sea-surface winds play an important role in heat transfer between the ocean and atmosphere and drive ocean circulation. This is particularly so in the Labrador Sea region. [Roy99c] used a Bayesian implementation of the hierarchical model discussed in Section 4 for modeling wind fields in the Labrador

Sea. That model is summarized here. The objective is to construct uniform wind and pressure fields from irregularly-spaced NASA scatterometer (NSCAT) wind data in the Labrador Sea. The fine resolution NSCAT[1] data provide the opportunity to resolve interesting features of wind fields which would ideally allow oceanographers to improve modeling of air–sea dynamics. Wind observations over the Labrador Sea for a single day are shown in Figure 1.

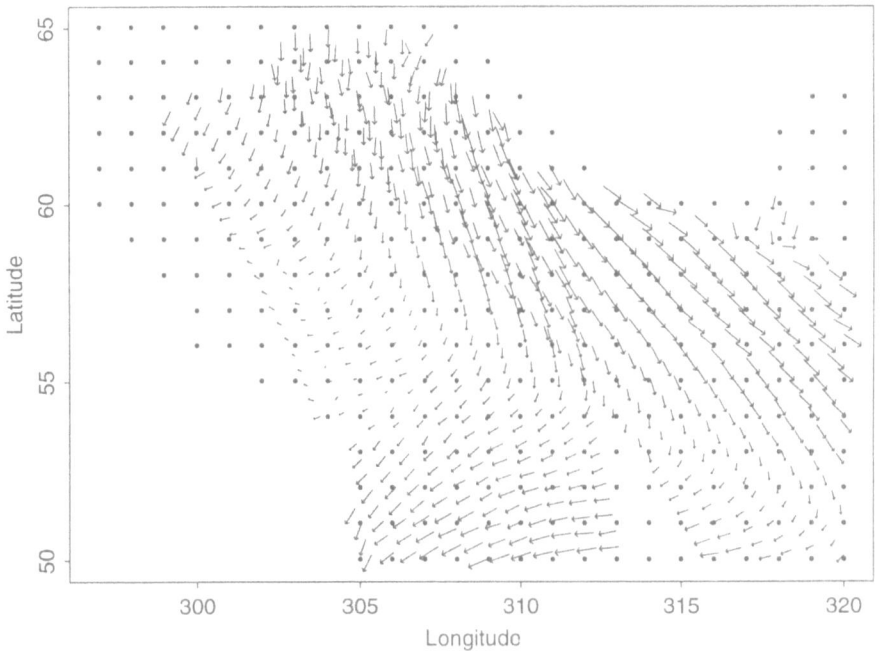

FIGURE 1. NSCAT observations over the Labrador Sea.

Wind is inherently bivariate, consisting of east–west and north–south magnitudes, denoted as $u(s)$ and $v(s)$, respectively. Furthermore, let $p(s)$ denote sea surface pressure at location s. Physics dictates that, as a first approximation, the wind vector at a location s is proportional to the pressure gradient; i.e., winds tend to blow along the pressure gradient (this is known as *geostrophy*). Other forces effect winds (e.g., friction), but the tendency of winds to behave geostrophically presents a good starting point for constructing a physically based statistical model.

[Roy99c] used a hierarchical model to quantify dependence of u and v on the pressure gradient. In this way, spatial dependence of the wind field was parameterized explicitly in a random field model for the pressure field. The key idea in this model is that, by conditioning on the spatially correlated pressure

---

[1] a satellite-borne instrument

field, the spatial structure in winds is much simplified. In contrast, to approach this problem in a cokriging context, one would be faced with modeling various covariance and cross-covariance functions (three of each).

Let $\mathbf{D}_u$ and $\mathbf{D}_v$ be $n \times 1$ data vectors of the u and v observations (irregularly spaced). Let $\mathbf{U}$, $\mathbf{V}$, and $\mathbf{P}$ be $N \times 1$ vectors of wind components and pressure on a grid consisting of N locations. Also, let $\mathbf{R}_\alpha = \text{Cov}(\mathbf{P}, \mathbf{P})$ be the variance-covariance matrix of the pressure vector. It is assumed that the covariance between pressure at any two points is parameterized by a covariance function depending on parameters α. The objective is prediction of $\mathbf{U}$ and $\mathbf{V}$ from the data $\mathbf{D}_u$, $\mathbf{D}_v$. In general, the pressure field may be observed along with the wind fields. In the present treatment, however, the pressure variable is treated as a latent process.

The first stage of the hierarchical model, the *data model*, relates wind observations to the gridded wind process:

$$\mathbf{D}_u, \mathbf{D}_v | \mathbf{U}, \mathbf{V}, \sigma_e^2 \sim \text{Gau}\left(\begin{pmatrix} \mathbf{HU} \\ \mathbf{HV} \end{pmatrix}, \begin{pmatrix} \sigma_e^2 \mathbf{I} & 0 \\ 0 & \sigma_e^2 \mathbf{I} \end{pmatrix}\right). \tag{3.16}$$

This model assumes that, conditional on $\mathbf{P}$, $\mathbf{U}$ and $\mathbf{V}$ are not spatially correlated. The relationship between gridded wind and the wind data is parameterized in the $n \times N$ matrix $\mathbf{H}$ that could be as simple as an incidence matrix relating observations to wind at the nearest grid point, or it could accommodate parameterizations which relate wind observations to weighted combinations of wind on nearby grid points.

The second stage of the model, the *process model*, relates the gridded wind process to the gridded pressure process and additional parameters:

$$\mathbf{U}, \mathbf{V} | \mathbf{P}, \mathbf{B}_u, \mathbf{B}_v, \Sigma \sim \text{Gau}\left(\begin{pmatrix} \mathbf{B}_u \mathbf{P} \\ \mathbf{B}_v \mathbf{P} \end{pmatrix}, \Sigma\right).$$

The $N \times N$ matrices $\mathbf{B}_u$ and $\mathbf{B}_v$ determine the conditional relationship between wind and pressure on the grid, and their precise form is discussed below. The random field model for the pressure field is given by

$$\mathbf{P} | \mu_p, \sigma_p^2, \mathbf{R}_\alpha \sim \text{Gau}(\mu_p \mathbf{1}, \sigma_p^2 \mathbf{R}_\alpha).$$

The covariance function of the pressure field was assumed to be the exponentially damped sinusoid (see [Thi85]):

$$\text{Cov}(p(s), p(s')) = [\cos(\alpha_2 ||s - s'||) + \frac{\alpha_1}{\alpha_2} \sin(\alpha_2 ||s - s'||)] e^{-\alpha_1 ||s - s'||}. \tag{3.17}$$

The final stage of the model included prior distributions on the various model parameters, including those in the matrices $\mathbf{B}_u$ and $\mathbf{B}_v$ and the covariance matrix for the pressure field.

An important feature of this model is that by parameterizing a meaningful relationship (in *mean*) between $(\mathbf{U}, \mathbf{V})$ and $\mathbf{P}$, the structure of Σ will be simplified. In [Roy99c], it was assumed that conditional on pressure, u and v are correlated at the same site, but not across sites. Thus, their model was

$$\Sigma = K_{2 \times 2} \otimes I_{n \times n},$$

where

$$K = \begin{pmatrix} \sigma_u^2 & \sigma_{uv} \\ \sigma_{uv} & \sigma_v^2 \end{pmatrix}.$$

The matrices B_u and B_v were parameterized according to geostrophy which has a straightforward difference representation on a grid. The matrices are functions of the 2×1 parameter vectors β_u and β_v, respectively and this parameterization is linear in β as discussed in Section 4.

In the analysis of [Roy99c], the model was fit using Markov chain Monte Carlo (MCMC) techniques. The posterior mean wind and pressure fields for a particular time are depicted in Figure 2.

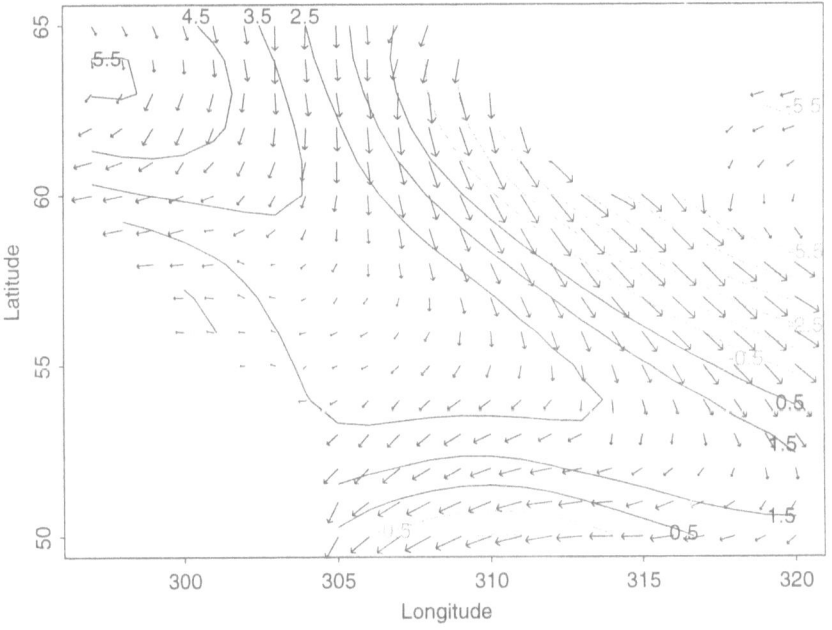

FIGURE 2. Posterior mean wind and pressure fields.

These results were particularly interesting because the posterior mean pressure field, estimated with *no* pressure data, resembled the European Center for Medium-Range Weather Forecasting's best guess at the true pressure field. See [Roy99c] and [Ber99c] for additional discussion and further details on the model and results.

The wind modeling problem is useful for studying the relationship between the hierarchical modeling approach of Section 4 and the cokriging approach of Section 2. The analysis cited above treats pressure as a latent process to enable modeling of the wind–pressure relationship in the conditional mean—a modeling strategy that is motivated by our understanding of the underlying physics governing the system. An equivalent model can be constructed that places this relationship in the second-moment structure of the model—i.e., the covariance and cross-covariance—instead of in the conditional mean. If we integrate the dependency of the unknown pressure variable from the model, we can explicitly compute the *implied* covariance structure of just the wind field under the geostrophic assumption. Indeed, this is precisely the approach taken by [Thi85] using the covariance function given by (3.17) to arrive at the joint covariance structure of wind fields to be used in the cokriging model specification. See [And98] for a similar development. Thus, the two models (hierarchical and cokriging) have a formal equivalence, suggesting that when known, conditional relationships exist, "proper" covariance structures can by deduced from these relationships. Moreover, this suggests a method for constructing interesting joint covariance models. That is, build a conditional model based on covariate information and examine the *unconditional* covariance structure implied by such models by integrating out these covariates.

It is interesting to note that the reverse approach of specifying the second moment structure, and then using this to exploit implied first-moment structure can also be useful. For example, this has been done by [Fed97] for constructing spatial designs, [Nyc99] for massive spatial prediction problems, and [Wik98b] for, at least in part, efficient Gibbs sampling of large spatial problems.

6.2 Modeling Temperature Fields

Another important bivariate field encountered in atmospheric science applications is temperature. Typically, daily or monthly temperature is summarized by its minimum and maximum, say $Y(s)$ and $Z(s)$. Unlike the previous example involving wind fields, there is not an obvious conditional model specification relating one of these variables to the other (except that $Z(s) > Y(s)$). And, although there are various fixed quantities such as latitude, elevation, time of year, etc., the two processes remain correlated, even conditional, on these covariates. Thus, if one's interest is in joint prediction of both fields, cokriging might seem the preferred method. It may be that conditional on wind, pressure, or other judiciously chosen atmospheric processes, a simplified model for minimum and maximum temperature may be constructed. Of course, this would depend on the particular application, data availability, and so forth.

One interesting observation regarding these two variables is that their spatial correlations appear to behave very similarly in many instances. It makes intuitive sense that if two sites have a highly correlated minimum temperature, they will have a similarly correlated maximum temperature. As an illustration of this tendency, we consider monthly minimum and maximum temperature from 94 U.S. Historical Climate Network sites in the Midwest shown in Figure 3. The pairwise correlations between sites for minimum and maximum temperatures, as functions of distances between sites, in the month of January, between 1920 and 1994, are shown in Figure 4. These two plots are similar, as are correlograms for other months of the year (omitted for obvious reasons of space). In fact, the correlations are nearly equal between any two sites for both minimum and maximum temperatures. Moreover, the cross-correlations between minimum and maximum temperatures (bottom panel of Figure 4) appear similar as well.

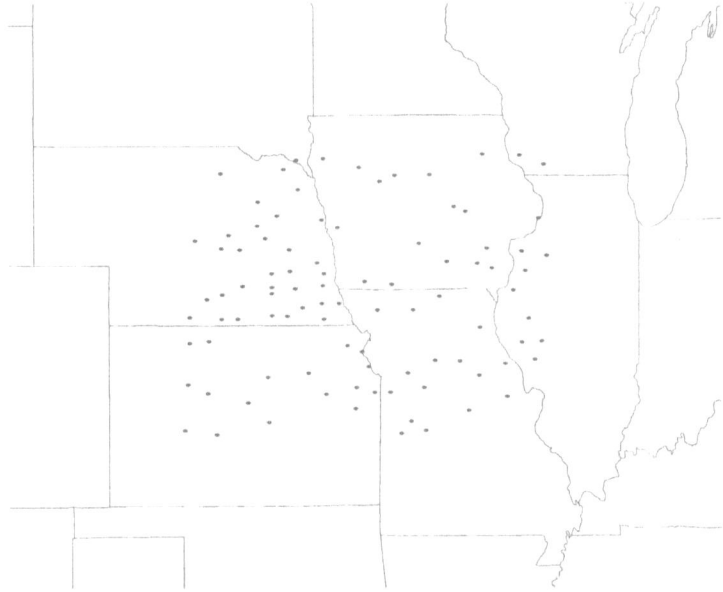

FIGURE 3. Locations of HCN temperature sites.

The consistency of the correlation patterns for the two variables suggests a latent process model under which both minimum and maximum temperatures have a common spatial process driving their variability. (This is not surprising since both of these variables are subject to the same climate forcings.) Consider the orthogonal transformation of Y and Z to *mean* and *range*, say U and V:

$$U(s) = \frac{1}{2}(Y(s) + Z(s)) \quad \text{and} \quad V(s) = Y(s) - Z(s).$$

The cross-covariance between these new variables is found to be greatly diminished which supports this latent process hypothesis. One reason is that this transformation removes the "common" correlation component from the cross-correlations of the new variables. The estimated correlations and cross-correlations of these two transformations for the same dataset are shown in

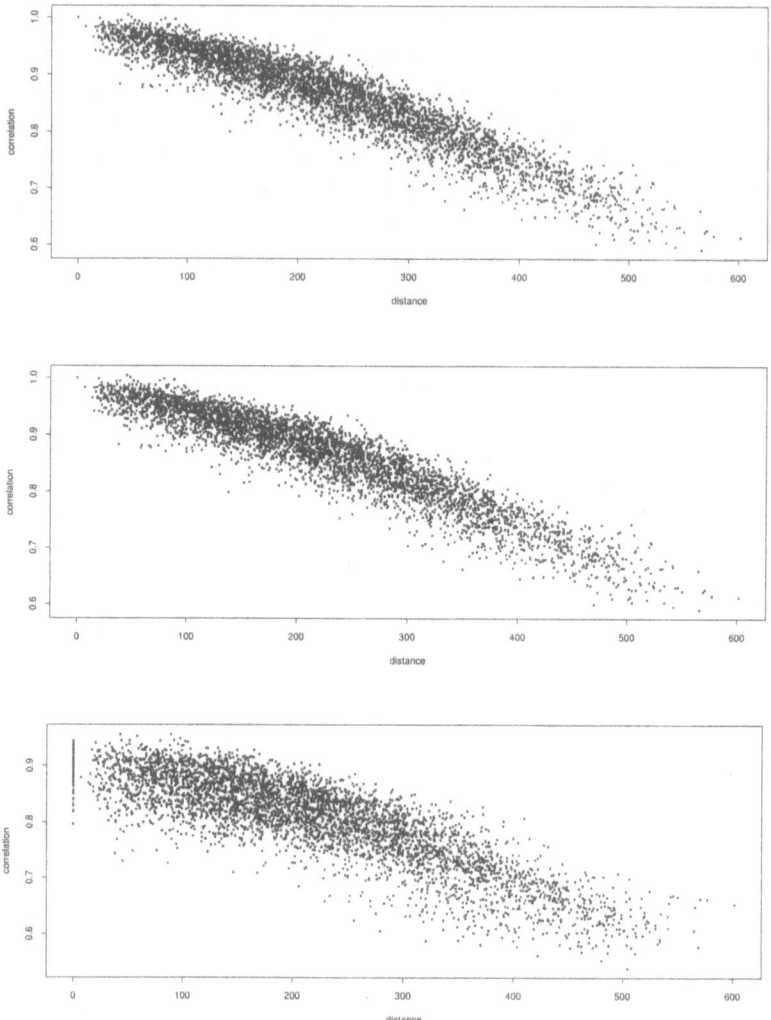

FIGURE 4. Correlations of January minimum (top panel) and maximum (middle panel). Cross-correlations are shown in the bottom panel.

Figure 5. Again, the marginal correlations look similar to one another, but perhaps with more of an exponential decay. It would be difficult to pose a model for the cross-correlations since they appear highly variable. Note that the correlations between one or more stations and the remainder appear quite different from the others (the values around −0.4 in Figure 5), perhaps suggesting some form of nonstationarity. Thus, it appears that the marginal covariance structure for each process has been simplified, and the cross-covariance structure has been reduced to a large extent. It is not clear, however, whether one can get by with a simpler (or no) model for the cross-covariance.

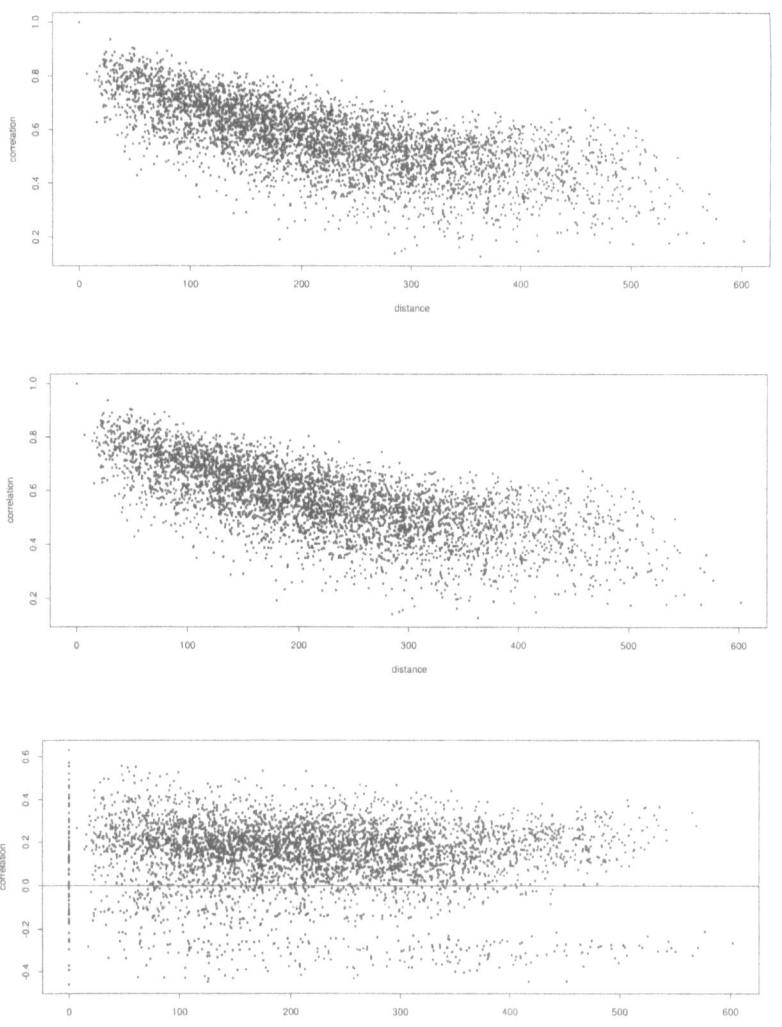

FIGURE 5. Correlations of January range (top panel) and mean (middle panel). Cross-correlations are shown in the bottom panel.

As an alternative modeling strategy, one might consider directly modeling a common latent process in a hierarchical model, perhaps with the latent process depending on other physical variables.

7   Conclusions

Multivariate spatial modeling techniques are important for studying atmospheric and environmental phenomena because processes interact with other processes. Typically, univariate analyses of spatial data are conducted not because it is believed that the process is independent of other processes but because:

1. data on related processes were not collected or are difficult to acquire;
2. the scientific questions to be examined are stated explicitly in terms of a single process; or,
3. incorporation of multivariate structures would greatly complicate the problem.

While adequate results might be attainable through univariate analyses, and certainly little can be done retrospectively to address item (1) above, a hierarchical modeling strategy can facilitate analyses by aiding in better statistical modeling of scientific understanding, and producing simpler statistical models through conditional parameterizations.

Acknowledgments

The author thanks former colleagues of the Geophysical Statistics Project for making the postdoc experience so enjoyable. Special thanks to Mark Berliner, Chris Wikle, Doug Nychka, and Wendy Meiring for reviewing drafts of this manuscript.

Hierarchical Space–Time Dynamic Models

Christopher K. Wikle
University of Missouri, Columbia, MO 65211, USA

1 Introduction

Virtually all atmospheric and oceanographic processes (e.g., wind, temperature, sea surface temperature, moisture) involve variability over space and time. For example, consider surface wind fields over the tropical oceans. Such fields are important factors in many processes that are of critical interest to the general public. These include tropical storm (hurricane) formation and maturation, and the development and strengthening of the El Niño–La Niña climate phenomena. One only need examine the governing partial differential equations for wind processes, or selected spatial and/or temporal averages of them, to see that mathematical and statistical descriptions of these dynamical processes depend on complicated temporal and spatial relationships. Furthermore, observations of geophysical processes typically include measurement errors and are often temporally and spatially incomplete. Both of these features can obscure the signal of interest.

To obtain "optimal" predictions in space and/or time, one must account for the stochastic nature of the process. In principle, given that one knows the complete space–time covariance structure of a process, it is a relatively simple matter to derive the BLUP. However, the space–time covariance structure of a geophysical process can be quite complicated, with spatial and temporal nonstationarities, anisotropy, and complex space–time interaction over many scales of variability. That is, space–time structure is often complicated by varying spatial behavior at different points in time, as well as varying temporal structure at different points in space. Furthermore, we seldom believe our knowledge of these complicated features is exact.

1.1 A Brief Review of Space–Time Modeling

There is a rich and growing literature on space–time modeling. Fundamentally, it is clear that in the absence of a temporal component, second-order "geostatistical" models can be used to represent spatial variability (e.g., intrinsically stationary error processes). These are *descriptive* in the sense that, although they model spatial correlation, there is no causative interpretation associated with them. Thus, for space–time modeling, the geostatistical paradigm assumes a descriptive structure for both space and time (i.e., covariance structures are directly specified). For example, one can extend the geostatistical "kriging" methodology for spatial processes by assuming that time is just another spatial dimension. Alternatively, one can treat time slices of a spatial field as variables and apply a multivariate or "cokriging" approach. Typically, simplifications such as separability between space and time are necessary to implement these approaches. See [Kyr99] for a comprehensive review of geostatistical space–time models. Although these approaches have been successful in many applications, there are fundamental differences between space and time, and it is not likely that realistic covariance structures can be specified that accurately capture the complicated dynamical processes as found in geophysical applications such as the tropical wind problem mentioned previously.

In the absence of a spatial component, there is a large class of time series models that could be used to represent the temporal variability (e.g., autoregressive error processes). These are *dynamic* in the sense that they exploit the fact that time flows in only one direction, and so the state of the process at the current time is related to what happened at previous times. Thus, one might consider the space–time process as a collection of spatially correlated time series in continuous space (e.g., [Ben79], [Rou90b], [Oeh93]) or on a spatial lattice (e.g., [Cli75], [Mar75], [Ali79], [Pfe80]). With the addition of a measurement error component in the model, the multivariate time–series models can be described through a state-space formulation. It is then natural to consider prediction via the Kalman filter (e.g., [Kal60a]). Generalizing the Kalman filter to a spatial context has its difficulties. Space does not have a natural ordering and hence the dynamic updating that is so important to Kalman filtering is missing. Yet, when there is also a temporal component, the spatial component can be used to index the state and measurement vectors. Examples of such models in the atmospheric and oceanic sciences include the Kalman filters that have been considered in data assimilation since the early 1980s (e.g., [Ghi81]) and are still very much the subject of research (e.g., [Mil97b]). Although these approaches include dynamical structures (i.e., they exploit the unidirectional flow of time and space–time interaction), without a descriptive spatial component one lacks the ability to perform spatial prediction at locations without observations.

1.2 Space–Time Dynamic Models

If both temporal and spatial components are present, it is natural to combine the temporally dynamic state-space approach and the spatially descriptive approach. We refer to such a model as a *space–time dynamic model*. These models have received much attention in recent years (e.g., [Cre96], [Gut94], [Goo94], [Hua96], [Wik96], [Wik97], [Mar98], [Mei98b], [Wik98a]). For a comparative discussion of these implementations, see [Cre98]. In general, these models assume that the observations are made up of the true state process plus measurement noise. Then, the state process is assumed to contain two space–time components, a spatial process evolving with time in a Markovian fashion, and a nondynamic space–time component to account for spatial structure that does not evolve with time. Often, large numbers of spatial observation and/or prediction locations lead to very high-dimensional models. In these cases, the state process can be projected onto orthogonal basis functions, reducing the dimensionality of the process [Wik97].

The methods listed above have proven extremely useful for providing good predictions of atmospheric and oceanic processes [Wik97], [Mei98b], [Wik99b]. Nevertheless, several objections can be raised. First, the estimation of parameters and associated uncertainty for these models can be difficult due to the curse of dimensionality. Second, these procedures are based soley on first- and second-moment properties of the process, thereby limiting their applicability. Furthermore, it is usually very difficult to include interrelationships between variables across both space and time. Finally, there is usually a substantial amount of scientific knowledge (i.e., physics) about the process of interest, which is not typically included in the above statistical formulations (with the notable exception of traditional Kalman filters in data assimilation, e.g., [Ghi81], [Mil97b]).

A Bayesian approach can overcome the limitations discussed above. Although Bayesian notions have been considered in the atmospheric and oceanic sciences for many years (e.g., [Eps85], [Tar87], [Lor86], [Cou97], [Ber99d]), fully Bayesian approaches in the context of space–time models are new. The advances and popularization of modern Bayesian computation (e.g., MCMC) methodologies in the 1980s have led to the plausibility of considering high-dimensional space–time problems from a hierarchical space–time perspective. These models allow flexibility in specifying space–time interactions and can include spatial and temporal nonstationarities, nonlinearity, and scientific knowledge.

In this chapter, we focus on hierarchical space–time dynamic models and the problem of predicting near-surface winds over a region of the tropical Pacific Ocean. The general ideas of hierarchical Bayesian space–time dynamic modeling are presented in Section 2. Section 3 contains a description of a "shallow water" dynamical system for winds and how this can be addressed through a combination of deterministic and stochastic approaches. The implementation of such a system with a "real-world" dataset is described in Section 4, followed by a brief discussion in Section 5.

2   Hierarchical Space–Time Dynamic Modeling

The essence of hierarchical modeling in the context of space–time processes is the organization of variables into three levels:

- *data*: observations (with noise) related to the process of interest, not necessarily at the same scale in space or time;
- *process*: the state variables of interest in our application; and
- *parameters*: statistical parameters and physical variables, all assumed to be random.

A space–time dynamic model will, in general, begin with a stage that considers measurement error in the observation of the true process via a data model (f_d) with associated parameters. This is followed by a process model that assumes the process at the current time t is a function of the process at previous times and some parameters plus instantaneous spatial noise. Finally, the parameters from the data and process models are given distributions at the next stage of the model hierarchy. The advantage of such a hierarchical approach is that elementary probability theory then suggests (via Bayes' theorem) the form of the distribution of the process and parameters given the data, the so-called posterior distribution:

$$[process, parameters| data] \propto [data|process, parameters]$$
$$\times [process|parameters][parameters],$$

where the bracket notation $[\cdot]$ represents a probability distribution.

2.1 A General Space–Time Dynamic Model

Assume that we are interested in some geophysical space–time "anomaly" (i.e., zero-mean) field $Y(\mathbf{s}; t)$ for $\mathbf{s} \in D$, where D is some spatial domain in Euclidean space and $t \in \mathcal{T}$, where $\mathcal{T} \equiv \{1, 2, \ldots, T\}$ is a discrete index of times. Further, assume we have noisy observations of the Y process, which we denote $Z(\mathbf{r}; t)$, for $\mathbf{r} \in D$. Our goal is to find the posterior distribution of $Y(\mathbf{s}; t)$, given the observations $Z(\mathbf{r}; t)$; $(\mathbf{r}; t) \in \mathcal{M}$, where $\mathcal{M} \subset \{D \cup \mathcal{T}\}$.

A mathematical formulation of the general space–time dynamic model can be expressed as

$$Z(\mathbf{r}; t) = f_d(Y(\mathbf{s}; t); \theta(\mathbf{r}, \mathbf{s}; t)) + \epsilon(\mathbf{r}; t), \tag{4.1}$$

$$Y(\mathbf{s}; t) = \int_{D_1} w_{\mathbf{s}}^{(1)}(\mathbf{u}; t-1) Y(\mathbf{u}; t-1) \, d\mathbf{u} + \cdots$$
$$+ \int_{D_p} w_{\mathbf{s}}^{(p)}(\mathbf{u}; t-p) Y(\mathbf{u}; t-p) \, d\mathbf{u} + \eta(\mathbf{s}; t), \tag{4.2}$$

$$distributions: \quad [w_{\mathbf{s}}^{(1)}(\mathbf{u}; t-1)] \ldots [w_{\mathbf{s}}^{(p)}(\mathbf{u}; t-p)][\epsilon(\mathbf{s}; t)][\eta(\mathbf{s}; t)], \tag{4.3}$$

where $f_d(\cdot)$ represents a relationship between the true process, Y, and the observation process, Z; $\epsilon(\mathbf{r}; t)$ is a measurement error process; $\eta(\mathbf{s}; t)$ is an instantaneous spatial noise process; $\theta(\mathbf{r}, \mathbf{s}; t)$ are parameters that relate the observation at location $\mathbf{r}$ to the process at location $\mathbf{s}$; $w_{\mathbf{s}}^{(j)}(\mathbf{u}; t - l)$ is a space–time interaction function at the jth lag; and $D_j \subset D$ is the spatial domain of interaction at the jth lag.

In the statistics literature, [Hua96] assume a simple measurement error model in the first stage such that $f_d(Y(\mathbf{s};t);\ \theta(\mathbf{r},\mathbf{s};t)) = Y(\mathbf{s};t)$ with $\epsilon(\mathbf{r};t)$ given to be white noise. Furthermore, they assume a stationary simultaneous noise structure η and simple structure on the interaction function, $w_{\mathbf{s}}^{(j)}(\mathbf{u};t-j) = \delta_{s,u}\alpha_j; j = 1,\ldots,p$, where $\delta_{s,u} = 1$ if $\mathbf{s} = \mathbf{u}$, and 0, otherwise. That is, they assume that the current value of the process only depends on past values at the same location. Since they assume that the "autoregressive" parameters α_j do not vary with prediction location, they are able to easily predict at locations for which observations are not available. However, such a model assumes implicitly a separable space–time interaction that may not be realistic for many climate processes. These authors choose to estimate the parameters rather than include the third stage (4.3). This so-called "empirical Bayesian" approach is easily formulated as a space–time Kalman filter for inference on the Y process (see Section 2.3).

[Gut94] and [Mei98b] consider a space–time dynamic model without the data (4.1) and the random parameter (4.3) stages. They assume nonstationary instantaneous spatial structure η and allow the interaction function to take the form $w_{\mathbf{s}}^{(j)}(\mathbf{u};t-j) = \delta_{s,u}\alpha_j(\mathbf{s}); j = 1,\ldots,p$. Similar to [Hua96], this model only assumes dynamic contributions from the prediction site at previous times. By allowing these "autoregressive" parameters to vary with space, they implicitly include nonseparable space–time interaction. However, since these interaction parameters are not random (i.e., no third stage in the hierarchy), it is difficult to estimate them at locations for which observations are not available, impeding off-site prediction.

[Wik97] assume a simple measurement error process as in [Hua96], but allow all spatial locations within a selected domain at the previous time to contribute to the evolution of the process. Specifically, they allow $w_{\mathbf{s}}^{(j)}(\mathbf{u};t-j) = w_{\mathbf{s}}^{(j)}(\mathbf{u}); j = 1,\ldots,p$, as well as spatially nonstationary instantaneous noise η. This model also includes a space–time process that does not evolve dynamically. The model is expressed in terms of a spectral decomposition on a complete and orthonormal basis set. They assume fixed parameters and get estimates via *method of moments* (MOM) for use in a space–time Kalman filter. The essence of this approach is discussed in Section 2.2. Note that a similar approach, but without the nondynamic space–time process, was independently investigated by [Goo94] and [Mar98].

A fully hierarchical implementation is demonstrated by [Wik98a]. In this case, the data model includes a function $f_d(\ )$ that maps prediction locations to observation locations, and includes white noise measurement error. This mapping at the first stage effectively discretizes the second stage and allows the replacement of the integrals in (4.2) with sums. Then, the interaction function $w_{\mathbf{s}}(\mathbf{u})$ takes a nearest neighbor structure (i.e., the process at a location at time t is related to the same site and its nearest neighbors at the previous time). Furthermore, the elements of $w_{\mathbf{s}}(\mathbf{u})$ are given spatial distributions at the third stage of the hierarchy. The advantages of this approach are that relatively complicated spatial structures on the interaction function parameters can alleviate the necessity for complicated instantaneous spatial structure in η, and the uncertainty in parameter estimation is explicitly considered.

Note that all of the space–time dynamic approaches discussed above have significant differences and complexities in their implementation. These differences are based on computational concerns as well as the goals of the desired inference. We have ignored these implementation details here. We have also omitted more complicated ARMA-like components to the process model (4.2) because such models are harder to justify physically for the geophysical processes of interest here.

2.2. Reformulated Space–Time Dynamic Model

For discussion, we consider the first two stages of the general space–time dynamic model (4.1) and (4.2). We further simplify these stages for ease of presentation:

$$Z(\mathbf{r};t) = \mathbf{k}_t'(\mathbf{r})\mathbf{Y}_t + \epsilon(\mathbf{r};t), \qquad (4.4)$$

$$Y(\mathbf{s};t) = \int_D w_\mathbf{s}(\mathbf{u};t-\tau)Y(\mathbf{u};t-\tau)\ d\mathbf{u} + \eta(\mathbf{s};t), \qquad (4.5)$$

where $\mathbf{Y}_t \equiv (Y(\mathbf{s}_1;t),\dots,Y(\mathbf{s}_n;t))'$ is the vector of Y at n prediction locations, and $\mathbf{k}_t(\mathbf{r})$ is the $n \times 1$ vector that maps the observation from location r to the prediction grid. With the goal of simplifying the integral formulation in (4.5) and reducing the dimensionality of the state process Y, we consider an equivalent spectral representation of the model (4.4) and (4.5). Similar to the approach of [Wik97], we let

$$Y(\mathbf{s};t) = \sum_{k=1}^M \phi_k(\mathbf{s})a_k(t) + \nu(\mathbf{s};t), \qquad (4.6)$$

$$w_\mathbf{s}(\mathbf{u};t-\tau) = \sum_{l=1}^M b_l(\mathbf{s};t-\tau)\phi_l(\mathbf{u}) + \gamma(\mathbf{s};t-\tau), \qquad (4.7)$$

where $\{\phi_k(\mathbf{s}), k = 1,\dots,M\}$ is a set of basis functions; $a_k(t)$ is the kth spectral expansion coefficient of the Y process at time t; $b_l(\mathbf{s};t-\tau)$ is the spectral expansion coefficient of the space–time interaction function; $\nu(\mathbf{s};t)$ is a "non-dynamic" (i.e., no temporal correlation) space–time process accounting for the truncation of the orthogonal series expansion of the Y process; and $\gamma(\mathbf{s};t-\tau)$ is the error associated with the truncation of the orthogonal series expansion of the interaction function. If the space–time interaction function $w_\mathbf{s}(\mathbf{u};t-\tau)$ is assumed to be fixed, then $\gamma(\mathbf{s};t-\tau)$ must be considered a deterministic error term and is often simply ignored in applications. Alternatively, one might consider the interaction function to be random with a mean given by the first term on the right-hand side of (4.7) and some *random* error structure γ. Such an approach gives a more realistic account of uncertainty in the model.

One can substitute (4.6) and (4.7) into (4.4) and (4.5) and, with appropriate independence assumptions, derive the equivalent model (written in vector form):

$$Z_t = K_t(\Phi a_t + \nu_t) + \epsilon_t, \tag{4.8}$$
$$a_t, = H_{t-\tau}a_{t-\tau} + \tilde{\eta}_t, \tag{4.9}$$

where Z_t is an $m \times 1$ vector of observations at time t; K_t is an $m \times n$ matrix that maps observation locations to prediction locations at time t; $\Phi \equiv [\phi_k(s_i)]_{i,k}$, $i = 1, \ldots, n$, $k = 1, \ldots, M$, is the $n \times M$ spectral basis function matrix; a_t is a $M \times 1$ vector of spectral coefficients; ν_t is an $n \times 1$ vector of the small-scale spatial process; ϵ_t is an $m \times 1$ vector of measurement error; $H_{t-\tau} \equiv JB_{t-\tau}$ is a $M \times M$ matrix of vector autoregression parameters; $J \equiv (\Phi'\Phi)^{-1}\Phi'$, $B_{t-\tau} \equiv [b_l(s_i; t-\tau)]_{i,l}$, $i = 1, \ldots, n$, $l = 1, \ldots, M$, is an $n \times M$ matrix of space–time interaction coefficients; and $\tilde{\eta}_t$ is a $M \times 1$ vector of spatially colored noise (in spectral space).

The model (4.8) and (4.9) is very familiar to atmospheric and oceanic researchers if ν is included with the measurement error, $H_{t-\tau}$ is time-invariant, and Φ is the empirical orthogonal function (EOF) basis set as derived through a Karhunen–Loéve expansion (e.g., [Pap65, pp. 457–461], [Wik96]). Recent examples include the sea level mapping of [Can96] and, implicitly, the principle oscillation pattern approach of [von95], and the so-called "linear inverse modeling" approach of [Pen93].

2.2.1   Empirical Bayesian Implementation

Rather than specify prior distributions on the parameters, we can take an empirical Bayes approach, whereby parameters are estimated via the data at hand. For state-space models, the empirical Bayesian model corresponds to the Kalman filter. If the parameter matrices in (4.8) and (4.9) are fixed and known, the optimal predictor of a_t, given data up to time t,

$$\hat{a}_{t|t} \equiv E(a_t | Z_t, \ldots, Z_1)$$

and associated prediction error covariance matrix,

$$P_{t|t} \equiv E\{(a_t - \hat{a}_{t|t})(a_t - \hat{a}_{t|t})' | Z_t, \ldots, Z_1\},$$

can be obtained via a recursive Kalman filter estimation approach (e.g., [Wik97]):

$$\hat{a}_{t|t} = \hat{a}_{t|t-1} + G_t(Z_t - K_t\Phi\hat{a}_{t|t-1}),$$
$$P_{t|t} = P_{t|t-1} - G_tK_t\Phi P_{t|t-1},$$

where the Kalman gain is

$$G_t = P_{t|t-1}\Phi'K_t'(\Sigma_\epsilon + \Sigma_\nu + K_t\Phi P_{t|t-1}\Phi'K_t')^{-1},$$

and

$$\hat{\mathbf{a}}_{t|t-1} = \mathbf{H}_{t-\tau}\hat{\mathbf{a}}_{t-1|t-1},$$
$$\mathbf{P}_{t|t-1} = \mathbf{H}_{t-\tau}\mathbf{P}_{t-1|t-1}\mathbf{H}'_{t-\tau} + \Sigma_\eta,$$

where $\Sigma_\epsilon \equiv \text{var}(\epsilon_t)$, $\Sigma_\nu \equiv \text{var}(\nu)$, and $\Sigma_\eta \equiv \text{var}(\tilde{\eta}_t)$. The optimal predictor for the process of interest, Y, is then easily derived (e.g., [Wik97]). For the case when the small-scale spatial noise process ν is assumed to be white noise (and included in the measurement error term, ϵ), this optimal predictor and associated prediction variance matrix is given by

$$\hat{Y}(\mathbf{s};t|t) = E[Y(\mathbf{s};t) \mid \mathbf{Z}_t, \ldots, \mathbf{Z}_1]$$
$$= \phi(\mathbf{s})'\hat{\mathbf{a}}_{t|t},$$
$$\sigma_Y^2(\mathbf{s};t|t) = \text{var}[Y(\mathbf{s};t) \mid \mathbf{Z}_t, \ldots, \mathbf{Z}_1]$$
$$= \phi(\mathbf{s})'\mathbf{P}_{t|t}\phi(\mathbf{s}) + \sigma_\epsilon^2,$$

where $\phi(\mathbf{s}) \equiv (\phi_1(\mathbf{s}), \ldots, \phi_M(\mathbf{s}))'$, and $\sigma_\epsilon^2 \equiv \text{var}(\nu(\mathbf{s};t))$. For the case when ν is not assumed to be white noise, see [Wik97]. In practice, the empirical Bayesian implementation of the Kalman filter requires that one must estimate the parameter and covariance functions from the data at hand. By substituting these estimates into the model, we no longer obtain exactly the conditional expectation nor the valid prediction error covariance. This is the case with the implementation of any BLUP methodology (e.g., kriging in space, and Kalman filters in time). Estimation of model parameters is typically accomplished through MOM approaches (e.g., [Wik97]) or by maximum likelihood (e.g., via the E–M algorithm, [Shu82]). Such estimation often assumes that the autoregressive parameter matrix in (4.9) be time-invariant since there usually is insufficient data to adequately estimate the time-varying parameters with these approaches.

2.3   A Bayesian Model

We now consider fully Bayesian extensions to the model given in (4.8) and (4.9) with a third stage analogous to (4.3) that allows for random parameters. Incorporating such randomness provides the opportunity for a comprehensive treatment of uncertainty about the model, and allows us to include prior information about the physics of the process. For simplicity in presentation, we temporarily assume that ν is spatial white noise and is combined with the measurement error ϵ term.

2.3.1   Data Model

Central to the Bayesian approach is that we can incorporate observational data at arbitrary scales in the estimation of the a process in space and time.

Then, we can formulate a probability model $[\mathbf{Z}_1, \ldots, \mathbf{Z}_T | \mathbf{a}_o, \mathbf{a}_1, \ldots, \mathbf{a}_T; \boldsymbol{\Sigma}_\epsilon, \boldsymbol{\Phi}]$. We will assume that

$$[\mathbf{Z}_1, \ldots, \mathbf{Z}_T | \mathbf{a}_o, \mathbf{a}_1, \ldots, \mathbf{a}_T; \boldsymbol{\Sigma}_\epsilon, \boldsymbol{\Phi}] = \prod_{t=1}^{T} [\mathbf{Z}_t | \mathbf{a}_t; \boldsymbol{\Sigma}_\epsilon, \boldsymbol{\Phi}]. \qquad (4.10)$$

That is, observations from one time period to the next are conditionally independent and only depend on the "current" value of the a process. To specify distributions and parameterizations for (4.10) we must consider what it is that the observations actually measure, what biases are present, and what is the precision of the measuring device? This leads to a functional form for the conditional distribution $[\mathbf{Z}_t | \mathbf{a}_t; \boldsymbol{\Sigma}_\epsilon, \boldsymbol{\Phi}]$.

2.3.2   Process Model

The space–time dynamic framework explicitly assumes a one-step, Markov vector autoregression (VAR) model for the evolution of the a process [Ber96], [Wik98a]. Thus, the joint distribution of the process is

$$[\mathbf{a}_o, \mathbf{a}_1, \ldots, \mathbf{a}_T | \mathbf{H}_t, \boldsymbol{\Sigma}_\eta] = [\mathbf{a}_o] \prod_{t=0}^{T-1} [\mathbf{a}_{t+1} | \mathbf{a}_t, \mathbf{H}_t, \boldsymbol{\Sigma}_\eta], \qquad (4.11)$$

where we have assumed that $\tau = 1$ in (4.8). Then, we must specify distributions for $[\mathbf{a}_0]$ and $[\mathbf{a}_{t+1} | \mathbf{a}_t, \mathbf{H}_t, \boldsymbol{\Sigma}_\eta]$.

2.3.3   Parameter Model

Finally, we must specify distributions for the parameters $[\mathbf{H}_0, \ldots, \mathbf{H}_{T-1}, \boldsymbol{\Sigma}_\epsilon, \boldsymbol{\Sigma}_\eta]$. Modeling assumptions are often made to simplify this distribution (e.g., VAR parameter matrices and variance–covariance matrices are conditionally independent). More importantly, it is at this stage that we can introduce much of our prior knowledge about the physics of the process. The distinction between process and parameters is somewhat arbitrary in this case. We will explore such ideas in Section 3.

3   Tropical Wind Process

For discussion, let us consider the case of wind fields in the tropics. It can be shown that if the depth of a rotating fluid (e.g., the atmosphere) is shallow compared to the horizontal extent, and the Coriolis acceleration is assumed to be a linear function of latitude, the dynamics that govern the fluid can be approximated by a linearized set of partial differential equations known as the "equatorial shallow water equations" (e.g., [Hol92, Sec. 11.4]):

$$\frac{\partial u}{\partial t} - \beta_0 yv + g\frac{\partial h}{\partial x} = 0, \tag{4.12}$$

$$\frac{\partial v}{\partial t} + \beta_0 yu + g\frac{\partial h}{\partial y} = 0, \tag{4.13}$$

$$\frac{\partial h}{\partial t} + \bar{h}\left(\frac{\partial u}{\partial x} + \frac{\partial v}{\partial y}\right) = 0, \tag{4.14}$$

where u, v are the east–west and north–south wind components, respectively; $\bar{h}$ is the mean fluid height; h is the difference between the total fluid depth and the mean fluid depth; g is the gravitational acceleration; and $\beta_0 = 2\Omega/r_e$, where Ω is the Earth's angular momentum and r_e the average radius of the Earth. Note that u, v, and h are all functions of spatial location $\mathbf{s} = (x, y)$ and time, t.

3.1 Deterministic View

For the moment, assume that we believe such a system is a reasonable approximation to the true dynamics of our system (e.g., surface winds in the equatorial Pacific Ocean). Define a state variable

$$\mathbf{Y}_t \equiv [u(\mathbf{s}_1, t), \dots, u(\mathbf{s}_n, t), v(\mathbf{s}_1, t), \dots, v(\mathbf{s}_n, t), h(\mathbf{s}_1, t), \dots, h(\mathbf{s}_n, t)]',$$

where $\{\mathbf{s}_1, \dots, \mathbf{s}_n\}$ are the spatial locations at which the process is considered (usually a "grid"). Applying a finite difference discretization to the time and space derivatives, this system could be written

$$\mathbf{Y}_{t+\tau} = \mathbf{H}\mathbf{Y}_t + \boldsymbol{\gamma}_t, \tag{4.15}$$

where $\boldsymbol{\gamma}_t$ represents the error introduced by discretization, and the matrix $\mathbf{H}$ is a sparse matrix made up of functions of the parameters β_0, $\bar{h}$, g, τ, and the grid spacing in the east–west (δ_x) and north–south (δ_y) directions. Then, given some initial condition $\mathbf{Y}_0$ and boundary conditions, predictions for all times and spatial locations follow naturally. This is a deterministic approach and, usually, no randomness is assumed. We could assume that $\boldsymbol{\gamma}_t$ is random and assign it a reasonable distribution. The choice of this distribution depends on the process, boundary conditions, and discretization procedure (see [Sci93] for an example). In the current shallow water example, such a model is probably not realistic for tropical wind systems unless $\boldsymbol{\gamma}_t$ is more than discretization error. That is, this process should include other physical features of the system (e.g., higher-order terms in the equations of motion).

3.2 Stochastic View

Now, consider the case where we have observations of u, v, and h, denoted $\mathbf{Z}_t^u, \mathbf{Z}_t^v, \mathbf{Z}_t^h$, for some set of locations $(\mathbf{r}; t) \subset M$ and time t. Our objective is to "predict" the true wind components at all locations and at all times. Specifically, we might be interested in the BLUP at some location $(\mathbf{s};t)$ given all available data. Thus, we assume that we can predict $u(\mathbf{s};t)$, $v(\mathbf{s};t)$, and $h(\mathbf{s};t)$ by some linear combination of the observations. For example,

$$\mathbf{Y}_t = \mathbf{\Lambda}_0\mathbf{Z}_t^w + \mathbf{\Lambda}_1\mathbf{Z}_{t-1}^w + \cdots + \mathbf{\Lambda}_{t-1}\mathbf{Z}_1^w, \qquad (4.16)$$

where $\mathbf{Z}_j^w \equiv (\mathbf{Z}_t^{\prime u}, \mathbf{Z}_t^{\prime v}, \mathbf{Z}_t^{\prime h})'$ is the set of m_j observations at time j. One could obtain the Λ's by minimizing the mean-squared error subject to unbiasedness constraints as is done in cokriging (see [Roy99a], this volume). However, in that case, it is clear that the parameters depend on the covariances and cross-covariances between u, v, and h at all time lags! The known class of valid space–time cross-covariance functions (i.e., ones that guarantee a positive definite joint variance–covariance matrix) that are manageable and realistic for such a multivariate space–time process is very small indeed! That is, even if one could identify a valid class of covariance functions, it would invariably be too simple to explain the complicated interactions of a system such as the tropical wind system described here. For example, to obtain the Λ's in (4.16) one might have to:

1. reduce the number of time lags involved in the prediction;

2. assume space–time separability; and/or

3. assume independence among the various random variables.

Any way you look at it, the task is generally daunting.

3.3   Combined Stochastic/Dynamic View

It was recognized early in atmospheric/oceanic data assimilation that one could use dynamical information (constraints) to help with the optimal interpolation (i.e., kriging) approach described in Section 3.2 (e.g., see [Dal91] for an overview). Although very useful, this approach also suffers from having to specify valid joint cross-covariance functions. Nevertheless, it is intuitively and scientifically appealing to make use of our knowledge of the physics (4.12)–(4.14) in a stochastic approach. We investigate such an approach from a hierarchical perspective.

3.3.1   Multivariate Hierarchical Space–Time Model

We do not actually believe that the simple system (4.12)–4.14) describes completely the wind process in the tropics. But, it is a good approximation and so we will decompose the state variables as a sum of an equatorial shallow water component (denoted by subscript "E") and some "correction" (denoted by subscript "c"):

$$\mathbf{u}_t = \mathbf{u}_{E,t} + \tilde{\mathbf{u}}_{c,t}, \tag{4.17}$$

$$\mathbf{v}_t = \mathbf{v}_{E,t} + \tilde{\mathbf{v}}_{c,t}, \tag{4.18}$$

$$\mathbf{h}_t = \mathbf{h}_{E,t} + \tilde{\mathbf{h}}_{c,t}. \tag{4.19}$$

Anticipating application, we will assume that h is not observed. We then consider the stochastic model

$$\begin{pmatrix} \mathbf{Z}_t^u \\ \mathbf{Z}_t^v \end{pmatrix} = \mathbf{K}_t \left[\begin{pmatrix} \mathbf{u}_{E,t} \\ \mathbf{v}_{E,t} \end{pmatrix} + \begin{pmatrix} \mathbf{u}_{c,t} \\ \mathbf{v}_{c,t} \end{pmatrix} \right] + \boldsymbol{\epsilon}_t, \tag{4.20}$$

$$\begin{pmatrix} \mathbf{u}_{E,t} \\ \mathbf{v}_{E,t} \end{pmatrix} = \mathbf{L}_{u,v} \begin{pmatrix} \mathbf{u}_{E,t-1} \\ \mathbf{v}_{E,t-1} \end{pmatrix} + \mathbf{L}_h \mathbf{h}_t, \tag{4.21}$$

where $\mathbf{K}_t$ is a sparse matrix that maps observations of the wind to the "true" wind process at prediction locations, and $\mathbf{L}_{u,v}, \mathbf{L}_h$ are the autoregression and regression matrices for the wind and height processes, respectively. We must then specify distributions for $\mathbf{K}_t$, $\mathbf{L}_{u,v}$, $\mathbf{L}_h$, $\mathbf{h}_t$, $\mathbf{u}_{c,t}$, $\mathbf{v}_{c,t}$, and $\boldsymbol{\epsilon}_t$. The $\mathbf{u}_{c,t}$ and $\mathbf{v}_{c,t}$ processes can be specified more completely by additional terms from the equations of motion, or they can be modeled as a correlated noise process. Because we don't actually observe h, it acts as a hidden (latent) process, and is represented as a random process at later stages of the hierarchy (e.g., see [Roy99c], [Roy99a]). One could assume that $\mathbf{L}_{u,v}$ and $\mathbf{L}_h$ are known, whereby we substitute their values from a deterministic finite difference discretization. However, such a discretization is an approximation and we shall account for this uncertainty by allowing these matrices to have random parameters.

The models given above can be simplified further if we return to the physics of the problem. In particular, we recognize that the system (4.12)–(4.14) has analytical wave solutions of the form

$$u_E^{l,p}(\mathbf{s}; t) = U_l(y) \cos(k_p x - \omega_{l,p} t), \tag{4.22}$$

$$v_E^{l,p}(\mathbf{s}; t) = V_l(y) \sin(k_p x - \omega_{l,p} t), \tag{4.23}$$

$$h_E^{l,p}(\mathbf{s}; t) = A_l(y) \cos(k_p x - \omega_{l,p} t), \tag{4.24}$$

where $u_E^{l,p}(\ )$, $v_E^{l,p}(\ )$, and $h_E^{l,p}(\ )$ represent the contribution to u_E, v_E, and h_E from the (l,p)th wave component; $k_p = 2\pi p/D_x$ where p is the east–west wavenumber and D_x is the east–west domain length; $\omega_{l,p}$ is the frequency of the (l,p)th wave mode; and $U_l(y)$, $V_l(y)$, $A_l(y)$ describe the north–south structure of the lth wave component (e.g., [Cus94, p. 287]). This north–south structure is proportional to Hermite polynomials that are exponentially damped away from the equator. For example,

$$V_l(y) = H_l(y^*) \exp(-0.5 y^{*2}), \tag{4.25}$$

where $H_l(\ )$ is the lth Hermite polynomial (with l corresponding to the number of nodes in the north–south direction); and y^* is the "normalized" latitudinal distance from the equator (i.e., $y^* = \beta_0 y/(\sqrt{gh_e}/\beta_0)^{0.5}$, where β_0 is a constant related to the ratio of the Earth's angular velocity to its radius, g is the gravitational acceleration, and h_e is the "equivalent depth" of the shallow fluid).

It is unrealistic to assume that real winds will propagate like perfect sinusoids as suggested in (4.22)–(4.24). To account for this uncertainty, as well as discretization approximations, we combine this physical model with a stochastic model. First, consider the v wind component and note that an elementary trigonometric identity allows us to rewrite (4.23) as

$$v_E^{l,p}(x, y; t) = \cos(\omega_{l,p}t)[V_l(y)\cos(k_p x)] + \sin(\omega_{l,p}t)[V_l(y)\sin(k_p x)]. \quad (4.26)$$

For each of our grid points $s_i \equiv (x_i, y_i), i = 1, \ldots, n$, we limit the number of equatorial wave solutions and modify this mathematical model to get

$$v_{E,t}(s_i) = \sum_{p=1}^{P}\sum_{l=0}^{L}\{a_{l,p;1}(t)[V_l(y_i)\cos(k_p x_i)] + a_{l,p;2}(t)[V_l(y_i)\sin(k_p x_i)]\}, \quad (4.27)$$

where $a_{l,p;1}(t)$, $a_{l,p;2}(t)$ are assumed to be random coefficients, with prior means as suggested by the cosine and sine terms in (4.26). Thus for all spatial locations on the prediction grid, we have

$$\mathbf{v}_{E,t} = \boldsymbol{\Phi}\mathbf{a}_t^v, \quad (4.28)$$

where $\mathbf{v}_{E,t}$ is the vector of v_E winds for all prediction grid locations at time t; and $\mathbf{a}_t^v$ is a vector of pairs of a's for each of the $J = P\times(L+1)$ combinations of p and l. The matrix $\boldsymbol{\Phi}$ is obtained by evaluating the shallow water basis functions at grid points. Specifically, for a total of J combinations, $\boldsymbol{\Phi}$ is an $n \times 2J$ matrix with columns $\phi_{2(j-1)+1}(x, y) = V_j(y)\cos(k_j x)$ and $\phi_{2(j-1)+2}(x, y) = V_j(y)\sin(k_j x)$ for $j = 1, \ldots, J$, evaluated at the coordinates of the n prediction grid locations.

As mentioned previously, the equatorial solutions are approximate. For example, we don't really expect that these waves will be present at all times, nor do we expect them to always propagate at the same frequency. Thus,

$$\begin{aligned}\mathbf{v}_t &= \mathbf{v}_{E,t} + \text{noise} \\ &= \boldsymbol{\Phi}(\tilde{\mathbf{a}}_t^v + \text{noise}) + \text{noise} \\ &\equiv \boldsymbol{\Phi}\mathbf{a}_t^v + \tilde{\mathbf{v}}_t, \quad (4.29)\end{aligned}$$

where we have assumed random parameters a^v rather than the deterministic parameters $\tilde{a}^v$, and we have added a stochastic noise term, denoted $\tilde{\mathbf{v}}_t$. Thus, we have a stochastic representation of the v wind component in terms of physically relevant basis functions. We can write similar stochastic, yet physically based, models for u and h.

Now, we can rewrite the measurement model (4.20) as

$$\mathbf{Z}_t^u = \mathbf{K}_t^u(\mathbf{\Phi a}_t^u + \tilde{\mathbf{u}}_t) + \boldsymbol{\epsilon}_t^u, \tag{4.30}$$

$$\mathbf{Z}_t^v = \mathbf{K}_t^u(\mathbf{\Phi a}_t^v + \tilde{\mathbf{v}}_t) + \boldsymbol{\epsilon}_t^v. \tag{4.31}$$

The Markovian process models for the equatorial wave components (4.21) can then be specified according to

$$\mathbf{a}_t^u = \mathbf{H}_u \mathbf{a}_t^u + \boldsymbol{\eta}_t^u, \tag{4.32}$$

$$\mathbf{a}_t^v = \mathbf{H}_v \mathbf{a}_t^v + \boldsymbol{\eta}_t^v, \tag{4.33}$$

where we have included random shocks $\boldsymbol{\eta}_t^u$, $\boldsymbol{\eta}_t^v$. We note the similarity between equations (4.30) (4.32) and (4.8) (4.9) given in the general space–time dynamic model overview, where the "tilde" processes in the wind model are analogous to the ν process in the general model.

Based on the deterministic physics, we might argue that $\mathbf{u}_{E,t}$, $\mathbf{v}_{E,t}$, and $\mathbf{h}_{E,t}$ are independent. Thus, the $\boldsymbol{\eta}^u$ and $\boldsymbol{\eta}^v$ processes are also independent and need not be dependent on $\mathbf{h}_t$. We make a similar assumption for the $\tilde{u}$ and $\tilde{v}$ processes. To complete the Bayesian formulation we must consider distributions for $\tilde{\mathbf{u}}_t$, $\tilde{\mathbf{v}}_t$, $\mathbf{H}_u$, $\mathbf{H}_v$, $\boldsymbol{\eta}_t^u$, $\boldsymbol{\eta}_t^v$, $\boldsymbol{\epsilon}_t^u$, and $\boldsymbol{\epsilon}_t^v$. These specifications will depend on the specific application.

3.4 Physically Informative Priors

Physical knowledge can be used to specify many of the prior distributions. For example, based on the derivation (4.22)–(4.27) we can specify the structure of the propagator matrices $\mathbf{H}_u$ and $\mathbf{H}_v$ (see [Wik99e]). Similarly, we can specify the prior structure of the innovation covariance matrices based on physical understanding. For example, in the case of tropical winds just above the surface, it has been shown [Wik99e] that the wind components exhibit a multiscale, "turbulent" scaling behavior in two dimensions. Specifically, the wind spectra follow a "$1/f$ process" whereby the energy spectrum is proportional to the inverse of the spatial frequency to some power

$$S_u(k_s) \propto \frac{\sigma_u^2}{|k_s|^\kappa}, \tag{4.34}$$

where $S_u(k_s)$ is the (spatial) power spectral density of u at wave number k_s, σ_u^2 is the variance of the u process, and κ is the spectral parameter (e.g., [Wor93]). In the tropics, κ has been found to be approximately 5/3 [Wik99d]. These ideas can be dated to [Kol41a], [Kol41b]. Because such processes often exhibit scale invariance, multiresolution bases such as wavelets are ideal for their modeling. Thus, we can model the $\tilde{u}$ and $\tilde{v}$ processes via a multiresolution wavelet process, where priors are placed on the variances so that they correspond to (4.34) (see [Ber99d] and [Wik99e] for more detail). Specifically, we let

$$\tilde{u}_t = \Psi b_t^u, \tag{4.35}$$
$$\tilde{v}_t = \Psi b_t^v, \tag{4.36}$$

where Ψ is an $n \times n$ matrix consisting of the two-dimensional multiresolution wavelet basis functions, and b_t^u, b_t^v are the associated spectral coefficients for $\tilde{u}_t$ and $\tilde{v}_t$, respectively. Then, it can be shown that under assumptions of spatial isotropy, the variance–covariance matrix of b_t^u and b_t^v can be considered diagonal with variances decreasing proportionally to scale according to

$$\sigma_b^2(l) \propto 2^{-l(1+\kappa)-1}, \tag{4.37}$$

where l is the level of the multiresolution decomposition (e.g., [Chi98]). Thus, we can use (4.37) to control the variability in the variance–covariance matrix of the wavelet coefficients to match roughly the turbulence scaling relationship (4.34). Note that we have implicitly assumed that conditional on the wavelet basis functions Ψ, the nonshallow water components of the wind are independent. Such an assumption is plausible, but difficult to prove based on the physics.

4   Ocean Wind Implementation

Consider the problem of near-surface wind fields over the tropical Pacific Ocean. There are no observations of such winds that are spatially and temporally complete at fine resolutions. Traditionally, the only wind observations available to scientists were relatively sparse buoy and ship observations, and the combined observation/deterministic model output produced by the large weather centers. These so-called "analysis" fields, although spatially and temporally complete (see Figure 1), are reported at relatively low spatial resolution (approximately 200 km) and they are fundamentally unable to resolve many of the small- to medium-scale features in the wind fields that are needed to understand the phenomena of interest (e.g., [Mil96], [Wik99d]). Alternatively, satellite-derived wind estimates from scatterometer instruments have become available in recent years. Although these estimates have high spatial resolution (e.g., 50 km), they cover limited areas at any given time (see Figure 1). Thus, our primary goal is to combine the analysis and satellite observations via the hierarchical space–time dynamic modeling strategy outlined above. We expect that by using the physics to develop our prior models, the posterior distribution of these winds will contain realistic physical signals over a range of spatial and temporal scales.

Satellite data from the NASA scatterometer (NSCAT) instrument is considered for the period from 28 October 1996 to 10 November 1996. Analysis

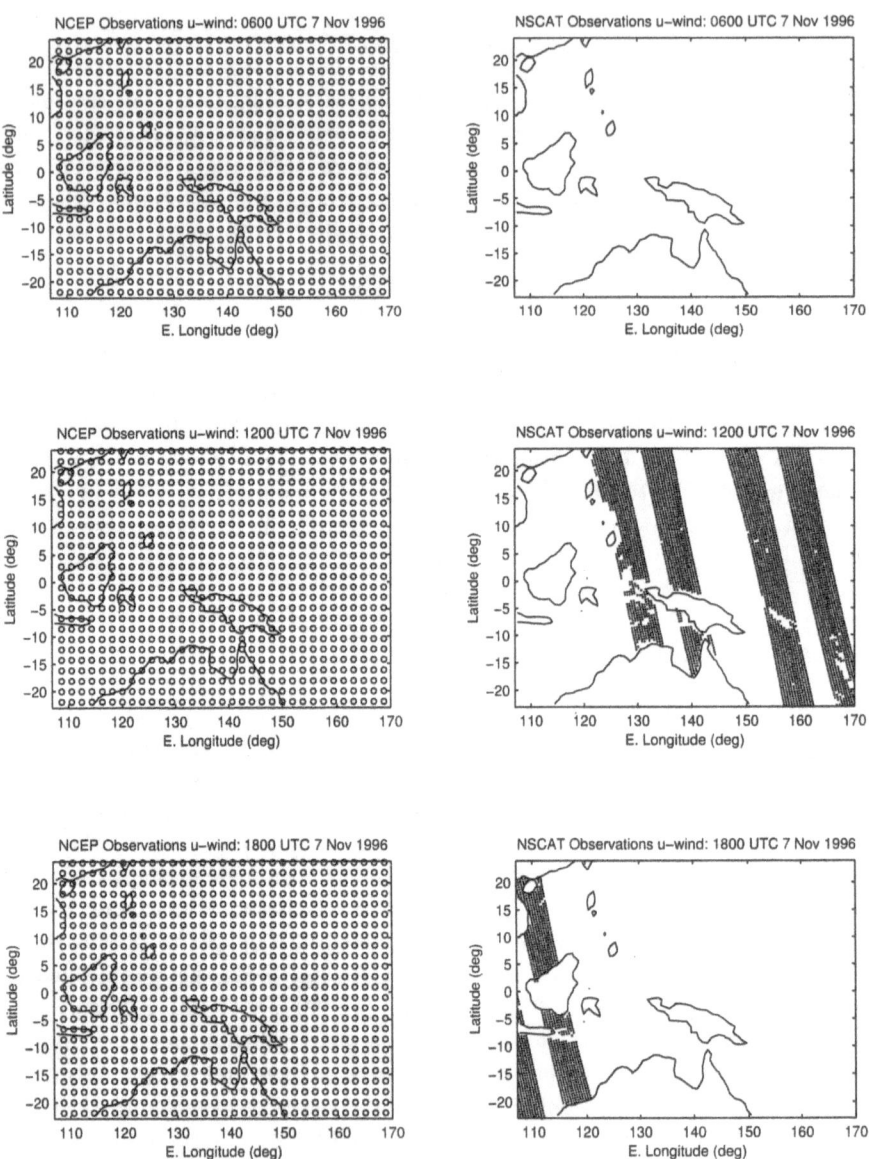

FIGURE 1. NCEP and NSCAT sampling locations within six hour time windows centered on 0600 UTC, 1200 UTC, and 1800 UTC on 7 November 1996.

winds for the same period are considered as produced by the National Centers for Environmental Prediction (NCEP). We consider these winds over the equatorial western Pacific Ocean, primarily because this is a critical region for many climate and weather phenomena. Specifically, we consider spatio-temporal predictions on a (64 × 48)-dimensional grid at one degree resolution in latitude and longitude and at six-hourly time intervals.

We selected two-dimensional wavelet basis functions (Ψ) from the Daubechies class with two vanishing moments, including a boundary modification at the edges (e.g., [Coh93]). The posterior distributions were found by MCMC via a Gibbs sampler. A general overview of Gibbs sampling can be found in [Gil96]. The specific details of our implementation can be found in [Wik99e].

4.1 Results

As a cursory examination of the success of our model, we examine aspects of the posterior distribution of the winds to check if they are realistic given our understanding of the physics. For example, Figure 2(a) shows the u-component of the wind from the NCEP analysis wind field at 1200 UTC on 7 November 1996. As mentioned previously, this field is known to be unrealistically smooth. Figure 2(b) shows the posterior mean of the shallow-water portion of our modeled winds for the same time. The posterior mean of the multiresolution wavelet portion of the modeled field is then shown in Figure 2(c). Finally, the complete blended v-component posterior mean field for this time is shown in Figure 2(d). It is apparent that this blended field contains much more information at intermediate and small scales than the analysis wind in Figure 2.

A very important characteristic of the wind field is its *divergence*. That is, the partial derivative of u with respect to the east–west direction, plus the partial of v with respect to the north–south direction. This scalar quantity measures the overall rate at which air is being transported away from a point along a horizontal plane (its negative, convergence, measures the rate at which air is being transported toward a point). It can be shown that convergence at the surface is related to convective storms in the tropics. Thus, as an independent check of the validity of our model, we can compare the convergence at the surface as derived from our posterior wind fields, to satellite images that are indicative of convective storms. Specifically, we compare to cloud top temperature data during a portion of the life-cycle of tropical cyclone Dale. In theory, the coldest cloud top temperatures should be roughly correlated to the strongest surface convergence. Figure 3(a) shows the wind and convergence field for the NCEP analysis winds, and Figure 3(b) shows the corresponding plot from the model posterior mean winds. Clearly there are more small and medium scale features in the convergence field for the modeled winds. Figure 3(c) shows the corresponding cloud top temperatures for this region and time. In general, the cloud structures evident in Figure 3(c) correspond

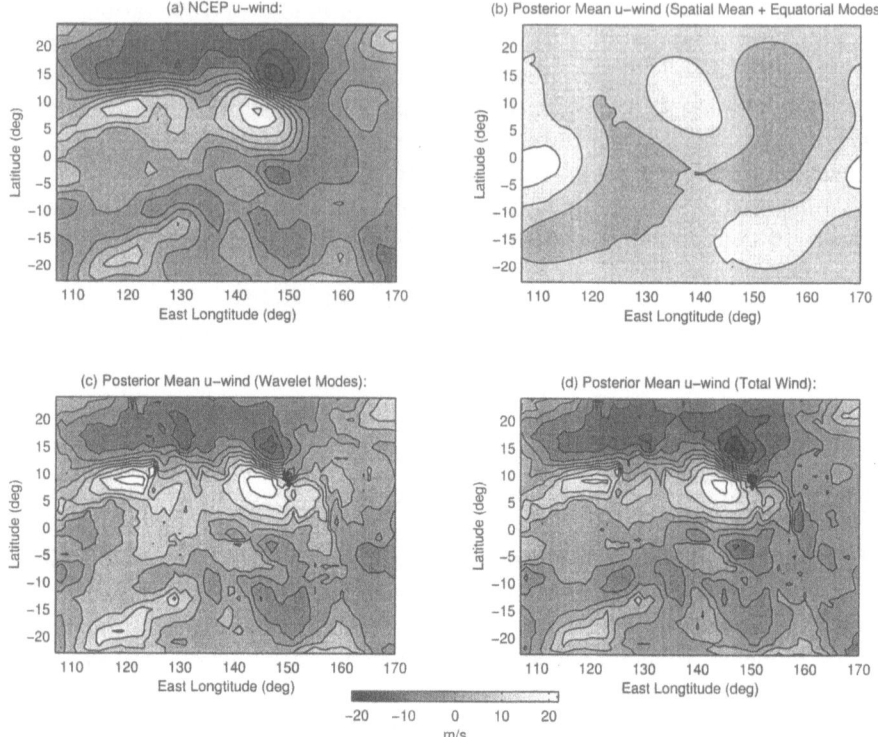

FIGURE 2. For 1200 UTC on 7 November 1996: (a) u-component of NCEP analysis wind field; (b) posterior mean u-wind component for equatorial shallow-water modes plus a spatial mean; (c) posterior mean of the wavelet modes; (d) total u-component posterior mean.

reasonably well with the modeled winds, and there is almost no correspondence between the analysis field and the clouds on any scales other than the very largest. This demonstrates that our modeled winds are physically quite realistic.

5   Discussion

This chapter presents a demonstration of how one can use dynamical information about a complicated process within the framework of a hierarchical space–time dynamic model. This was illustrated with a relatively simple *linear* set of partial differential equations. For most processes in the atmospheric/oceanic sciences, we must consider nonlinear interactions as well (e.g., [Sci93]). Similarly, we have made simplistic assumptions about the interaction of small scale processes in our description. More complicated models for such interactions can

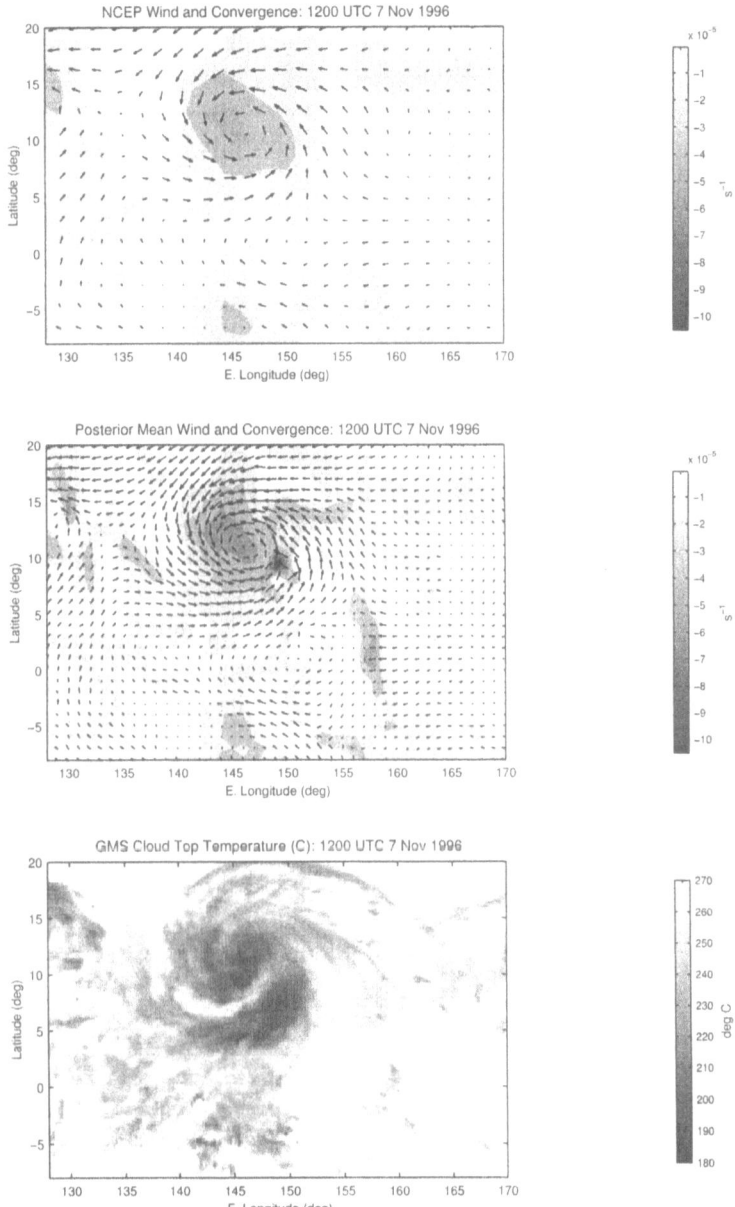

FIGURE 3. For 1200 UTC on 7 November 1996: (a) convergence and wind fields from the NCEP analysis; (b) posterior mean convergence and wind field from the hierarchical model; (c) cloud top temperatures.

be incorporated in the hierarchical framework.

It is somewhat unnatural to consider the set of partial differential equations as difference equations as we described. A more realistic approach would be to consider the process from a continuous time perspective. In that case, the time evolution of the process is described in a probabilistic framework through a Fokker–Planck equation. A simple example from a Bayesian perspective is discussed in [Ber96].

Finally, a significant problem in deterministic modeling of geophysical systems is the coupling of models for different processes. For example, in climatology, models for the atmosphere and ocean must be "coupled" to allow feedback between the processes. The hierarchical framework provides a natural approach for such coupling.

Acknowledgments

The author would like to thank L.M. Berliner, D. Nychka, and J.A. Royle for valuable comments on an early draft.

Experimental Design for Spatial and Adaptive Observations

Zhan-Qian Lu
Data Analysis Product Division, Mathsoft, Inc., Seattle, WA 98109, USA

L. Mark Berliner
Ohio State University, Columbus, OH 43210, USA

Chris Snyder
National Center for Atmospheric Research, Boulder, CO 80307, USA

1 Introduction

Global numerical weather prediction (NWP) uses large amounts of routinely-collected observations of the atmosphere. Recently, there has been substantial interest in augmenting these routine observations with specially designed additional observations. The idea is that these new observations would be chosen *adaptively* in an effort to *target* particular spatial regions (or variables) whose observation would be especially useful in reducing errors in forecasts. These observations could be used for tracking and forecasting small-scale and short-term phenomena such as fronts and storms. Indeed, based on current estimates of the state of the atmosphere, physical reasoning, and past experience, meteorologists would want to target particular locations whose information could help to either predict the development of a storm or predict a storm's path or track. The enterprise is made feasible with the availability of mobile observing platforms, in particular, aircraft. This general area is known in meteorology by the phrases *adaptive, targeted,* or *supplemental observations*. A recent multinational experiment, FASTEX, was a major stimulus for a surge of research along these lines. Recent references include [Jol97], [Lor98], [Pal98], and [Bis99].

The problem of adaptive observations falls within the general rubric of statistical experimental design. However, much of the statistical literature focuses on experimental design in support of system identification or estimation, rather than for prediction. Further, experimental design in the context of spatial or spatio–temporal processes is not discussed in most experimental design texts. Hence, we investigate general formulations of optimal statistical design for sequential prediction problems for complex spatial or space–time processes. Such problems arise not only in NWP, but also in many environmental science contexts.

We formulate the adaptive observations problem mathematically as one of optimal design for observing a high-dimensional dynamical system, with the goal of best estimation of the state vector at the instant the adaptive observations are collected or of best prediction of the state at some future time. In both cases "best" refers to minimization of some expected loss measure. Here, we focus on mean-squared estimation (or prediction) error. Consider the following idealized version of the adaptive observations problem. Let X be an n-dimensional vector representing the state of the atmosphere. In particular, X is composed of the discretized variables used in an NWP computer model. Let X_0, X_1, and X_2 be the states of the atmosphere at times, t_0, t_1, and t_2, respectively. The adaptive observations problem can be stated as:

> Using all available data and information at $t = t_0$, select additional observations to be taken at $t = t_1$ to optimize either a mean-squared error criterion for the estimation of X_1 or a mean-squared prediction error criterion for X_2.

A simple view of this description is to imagine that at t_0, we must file the flight plan for an aircraft that will take observations 24 hours from now (t_1), with the intent of obtaining either best estimation of the state then or best prediction, based on all information available at time t_1, of the state two to three days from now (t_2).

The choice of design ought to depend on our best knowledge of the atmospheric states X_1 or X_2. Following the methods of modern NWP, our analyses are based on combinations of:

1. information contained in all available data; and
2. knowledge about the dynamical evolution of the state or flow of the atmosphere.

The second information source is represented via a numerical model. To conveniently formalize the incorporation of these information sources, a Bayesian approach is used. Note that since both the quantities to be estimated or predicted and the information sources about them are time-dependent, so are the adaptive designs.

In the context of NWPs, major difficulties in implementation arise due to the strongly nonlinear, chaotic nature of the underlying weather process and the high dimensionality of the system. Both of these issues moderate the development of exact optimal procedures, mandating the use of approximations. In particular, we base our analyses on *linearizations* of the nonlinear dynamical models. (Linearization is common in NWPs.) This is clarified in Section 3.

In the sequel of this section, we offer two very brief reviews. The first provides some background concerning NWPs. Section 1.2 includes some references and general remarks concerning statistical experimental design. An approach to Bayesian design for a static, purely spatial random process, along with results of interest in adaptive observations, are presented in Section 2. Section 3 considers the design of data collection in support of the prediction of general dynamical processes. We relate the formalizations of Section 2 directly to the adaptive observations problem. We conclude the chapter with a discussion of some open problems of interest to both statisticians and meteorology researchers. Finally, as mentioned earlier, there already exists an important literature on adaptive observations. However, for brevity, we do not provide comparisons of the results of our approach with others. Such comparisons can be found in [Ber99b].

1.1   Scientific Background

Operational NWPs, as practiced at major weather prediction centers such as the National Centers for Environmental Prediction, involves a data assimilation process in which vast amounts data are processed to produce estimates or *analyses* of the state of the atmosphere as some time. These analyses are used to specify initial conditions such as input to a numerical model which leads to predictions [Dal91]. However, despite great advances in physical modeling, computational resources, and enhanced observational assets, there are intrinsic limits to:

1. how well one can know the current state of the atmosphere; and
2. how far ahead one can reasonably predict the weather.

The range or horizon of skillful predictability varies, depending on the specific features to be predicted and, of course, the method of prediction. For the scales of motion resolved by present global NWP models, it is generally believed that the range of predictability is limited to no more than two weeks. Note that the skill of predictions involve measuring their added value beyond obvious seasonal and climatological adjustments.

The predictive power of the models used in global NWP depend in great measure on their resolution and the lead-time of the forecast. For example, a global model with resolution of about 100 km at the surface is subject to an "error doubling effect" within about 36 hours. To clarify this, consider a specified value of the current state and a perturbation of that value. If these values are each substituted into the same NWP model to compute their 36 hour forecasts, the resulting predicted values will typically differ by twice the magnitude of the original perturbation. Fundamentally, this is due to the fact that the numerical model representation of weather dynamics is a chaotic process [Lor93].

In summary, errors in point-forecasts based on NWPs generally arise from three sources:

- Analysis errors in estimates of current weather conditions. These arise from a variety of causes including natural measurement error in the data; incomplete sampling of the atmosphere; uncertainties associated with the combination of heterogeneous sources of data; and their processing in numerical models.

- Error doubling effects due to the nonlinear (chaotic) nature of the dynamics.

- Forecast model error. The forecast models used in NWPs are of course imperfect—even if the present state of the atmosphere were known precisely, forecasts of future states would be in error. Forecast model errors are associated with inherent space–time discretizations of the governing partial differential equations. In particular, many processes with scales smaller than that of the discretization (cumulus clouds, for example) are either overlooked or only crudely represented in present forecast models.

To account for the uncertainty in the state of the atmosphere, it is convenient to treat the state as a random quantity. Knowledge about it is summarized in a probability distribution. This is in direct correspondence with the Bayesian viewpoint [Ber85], [Ber99c]. See [Lor86] and [Tar87] for a discussion in the context of data assimilation and NWP.

1.2   Overview of Statistical Design

R.A. Fisher blazed the trail for statistical experimental design while working at the Rothamsted Experimental Station, England, in the early 1920s. Fisher's interest was in the optimal arrangements of treatments (fertilizers) and blocks (spatial locations), so the interactions between treatments (fertilizers) and soil properties would not prohibit learning about the comparative values of fertilizers.

In the 1950s, G.E.P. Box applied statistical experimental design to industrial and manufacturing concerns [Box69], [Box78]. In Japan, G. Taguchi advocated the application of experimental design in support of the specification of manufacturing processes that would yield high quality products. Part of Taguchi's method seeks to reduce the effect of noise in inputs by taking advantage of nonlinearities of the underlying process.

The design problem was treated in a formal, decision-theoretic framework by J. Kiefer and V. Fedorov. The result is now known as *optimal design theory*. Depending on a statistical criterion, an optimal design is obtained as the solution of a mathematical optimization problem. See [Fed72] and [Sil80]. A review of Bayesian experimental design is given in [Cha95].

Sequential statistical inference was developed by A. Wald in the 1940s. The resulting theory of *sequential analysis* has seen rapid development and various applications. Sequential experimental design has been applied to deal with various problems including nonlinear models [Che72], [Tit80].

A general account of spatial statistics, including an introduction to spatial design, is given in [Cre93]. The design of environmental monitoring networks is discussed in [Swi79], [Cas84], [Le,94], and [Nyc96]. [Rod74] and [Wik99c] consider space–time issues in designing environmental monitoring networks. The application of spatial design in selecting inputs in computer experiments is discussed in [Sac89].

2   Experimental Design: Spatial Fields

Consider an n-dimensional random vector $\mathbf{X}$, having mean $\boldsymbol{\mu}$ and covariance matrix B. Examples include gridded values of some variable defined over a spatial region; that is, the index of the elements of $\mathbf{X}$ actually represents gridded spatial location. Both $\boldsymbol{\mu}$ and B are assumed known. The covariance matrix B describes spatial covariances among the values of the X variable at different sites. Suppose our observations, a d-dimensional vector $\mathbf{Y}$, follow the statistical model or *observation equation*

$$\mathbf{Y} = K\mathbf{X} + \varepsilon \qquad \text{where} \quad \varepsilon \sim N(0, \Sigma). \tag{5.1}$$

Here K is a $d \times n$ *design matrix*. The design problem involves the choice of optimal K. The matrix Σ describes the variability of *measurement errors* associated with the measuring instrument to be used. We note that, in general, Σ depends on K, but we do not reflect this in our notation. Also, the observation equation (5.1) is often based on a linearized approximation. That is, the mean of the data may actually be a nonlinear function of the state $\mathbf{X}$.

Assume that

$$\mathbf{X} \sim N(\boldsymbol{\mu}, B). \tag{5.2}$$

Bayes' theorem provides a recipe for updating our *prior* information about $\mathbf{X}$, as reflected in (5.2), in light of the observed data values $\mathbf{Y} = \mathbf{y}$, leading to a *posterior distribution* for $\mathbf{X}$:

$$\mathbf{X}|\mathbf{Y} = \mathbf{y} \sim N(\boldsymbol{\nu}, A), \tag{5.3}$$

where

$$\boldsymbol{\nu} = \boldsymbol{\mu} + BK^T(\Sigma + KBK^T)^{-1}(\mathbf{y} - K\boldsymbol{\mu}), \tag{5.4}$$
$$A = B - BK^T(\Sigma + KBK^T)^{-1}KB, \tag{5.5}$$

or, equivalently,

$$A = (B^{-1} + K^T\Sigma^{-1}K)^{-1}. \tag{5.6}$$

Remarks

1. The results in (5.5) and (5.5) are well known in the data assimilation literature, as is their derivation via Bayes' theorem [Lor86], [Cou97]. Nevertheless, most writings in data assimilation provide an alternate derivation, known as three-dimensional variational data assimilation (3-D Var). ("Three" here refers

to three spatial dimensions.) The approach is to find that value $\mathbf{x}$ to minimize the quantity

$$J(\mathbf{x}) = (\mathbf{x} - \boldsymbol{\mu})^T B^{-1}(\mathbf{x} - \boldsymbol{\mu}) + (\mathbf{y} - K\mathbf{x})^T \Sigma^{-1}(\mathbf{y} - K\mathbf{x}). \qquad (5.7)$$

This alternative approach is readily recognized by statisticians as one using penalized least squares or log-likelihood arguments, or alternatively, computation of a posterior mode in a Bayesian context.

It is easy to optimize (5.7) by setting its vector of first derivatives to zeros:

$$B^{-1}\mathbf{x} + K^T \Sigma^{-1} K\mathbf{x} - B^{-1}\boldsymbol{\mu} - K^T \Sigma^{-1}\mathbf{y} = \mathbf{0},$$

yielding the solution

$$\hat{\mathbf{x}} = (B^{-1} + K^T \Sigma^{-1} K)^{-1}(B^{-1}\boldsymbol{\mu} + K^T \Sigma^{-1}\mathbf{y}). \qquad (5.8)$$

This result coincides with $\boldsymbol{\nu}$ in (5.5). This must be the case, since the posterior mode and mean coincide under the normality assumptions used here.

2. The derivations and results (5.5) and (5.5) are also recognizable as cousins of *kriging* with measurement error procedures [Cre93].

3. There will typically be constraints on the form of design matrices K. One natural constraint is that the admissible K be *incidence matrices*. Namely, each row of K is a vector of $(n - 1)$ zeros and a single 1, so that a row indicates a particular coordinate of $\mathbf{X}$ to be targeted. Also, physical constraints such as the spatial contiguity of sites observed significantly reduce the number and forms of admissible designs.

4. Note that we do not require that the rows of K must be different. That is, the optimal design may include multiple observations of the same linear combination of $\mathbf{X}$.

2.1   Optimal Design

The proposition of an optimal design problem begins with a criterion by which designs may be compared. If our ultimate problem is point prediction of the state $\mathbf{X}$, we need a criterion by which we judge that prediction. Here, we adopt the "mean-squared prediction error" as our criterion [Ber99b].

A standard result from prediction theory suggests that under that criterion, our point prediction should be the posterior expectation of $\mathbf{X}|\mathbf{Y} = \mathbf{y}$; in our setting, $\boldsymbol{\nu}$. The corresponding measure of prediction error is A, the predictive covariance matrix. In general, A is a function of the yet-to-be-observed data $\mathbf{Y}$; hence, we could only optimize the expected value of A. However, in the formulation given here, A does not depend on $\mathbf{Y}$, and hence, we can optimize it directly (this fact strongly relies on the normal distribution assumptions).

Since A is a matrix, it makes no sense to suggest finding a K which "minimizes" A. Optimal design theory focuses on minimizing a selected scalar-valued function of A. The choice of that function depends on the goals of the prediction analysis. Some standard criteria are discussed next.

A-optimality. An *A-optimal* design minimizes the average (or total) mean-squared prediction error. In our case, we are to minimize $\text{tr}(A)$. By (5.5), we are to minimize

$$\text{tr}(B - BK^T(\Sigma + KBK^T)^{-1}KB),$$

or, equivalently, maximize

$$\text{tr}\{(\Sigma + KBK^T)^{-1}KB^2K^T\} \tag{5.9}$$

over K.

D-optimality. A *D-optimal* design minimizes the volume of prediction regions, computed by finding ellipsoidal regions in $\mathcal{R}^n$ having a fixed probability content, based on the posterior distribution (5.3). This turns out to be equivalent to minimizing the determinant of A, or equivalently, maximizing

$$\det(A^{-1}) = \det(B^{-1} + K^T\Sigma^{-1}K). \tag{5.10}$$

We can also write

$$\det(A) = \det(B)\det(\Sigma)/\det(\Sigma + KBK^T). \tag{5.11}$$

Hence, a D-optimal design, if $\det(\Sigma)$ is assumed to be the same across different locations, is found by maximizing

$$\det(\Sigma + KBK^T) \tag{5.12}$$

over K.

E-optimality. An *E-optimal* design minimizes the largest prediction error of any (normalized) linear combination of $\mathbf{X}$. (That is, minimize the largest variance of $\mathbf{c}^T\mathbf{X}$ over all $\mathbf{c}$ such that $\mathbf{c}^T\mathbf{c} = 1$.) Such designs are obtained by minimizing the largest eigenvalue of A. Equivalently, an E-optimal design is found by searching over K so that the minimum eigenvalue of

$$A^{-1} = B^{-1} + K^T\Sigma^{-1}K \tag{5.13}$$

is maximized.

A variety of other criteria, interrelationships, and generalizations have been studied; see [Puk93] for a discussion.

In some applications, only predictions of a subset of $\mathbf{X}$ may be of interest. Criteria and corresponding optimal designs tailored for such *local* prediction problems can be defined easily by applying the above definitions to the appropriate submatrix of A.

2.2 Special Solutions

Selection of d sites. We now consider the reduction to incidence matrices as admissible K in some detail. The design problem is to select exactly d "sites," or elements of $\mathbf{X}$, for observation. A useful representation of the design is as a list of sites, say $s_d = \{i_1, i_2, \ldots, i_d\}$, where $s_d \subset \{1, 2, \ldots, n\}$.

A-optimality. An A-optimal design maximizes

$$\mathrm{tr}\{(\Sigma + B[s_d, s_d])^{-1} B[s_d, .]B[., s_d]\}, \tag{5.14}$$

where $B[s_d, s_d]$ denotes the submatrix of B consisting of rows and columns indexed by s_d and $B[s_d, .]$, or $B[., s_d]$ is the submatrix consisting of rows or columns indexed by s_d.

Suppose $d = 1$. Then Σ is a scalar, say σ_i^2. Let b_{ij} denote the (i, j)th element of B. An A-optimal design is given by a site i for which

$$\frac{\sum_{j=1}^n b_{ij}^2}{\sigma_i^2 + b_{ii}} \tag{5.15}$$

is maximized.

An ad hoc suggestion for targeting with $d = 1$ is to observe at the site having maximum prior variance; i.e., the largest diagonal element of B. While the diagonal elements of B are critical in A-optimality, they need not dominate the design; (5.15) suggests that both variances and covariances among sites are important. The intuition is that information about some sites can provide enough information about other sites to overwhelm consideration solely of the marginal variance. As a simple example, suppose

$$B = \begin{pmatrix} 1.0 & 0.0 & 0.0 \\ 0.0 & 0.9 & c \\ 0.0 & c & 0.8 \end{pmatrix},$$

where $c^2 < 0.72$. Further, suppose that all $\sigma_i^2 = \sigma^2$. First, one can show using (5.15) that Site 2 is preferred to Site 3 for all σ^2. (This makes sense since, by definition, Sites 2 and 3 are equally correlated, yet not correlated with any other sites. Therefore, in choosing between them, we should choose that site with largest variance.) Next, Site 1 is preferred to Site 2 whenever

$$\frac{0.09 + 0.19\sigma^2}{1 + \sigma^2} > c^2.$$

Algebraic analysis of the inequality reveals that:

1. Site 1 is preferred to Site 2 whenever $c^2 < 0.09$; and
2. Site 2 is better whenever $c^2 > 0.19$.

For intermediate cases the inequality is easy to check.

D-optimality. If $\det(\Sigma)$ is constant for all designs, a D-optimal choice of the sites s_d is found by maximizing (see (5.12))

$$\det(\Sigma + B[s_d, s_d]).$$

If $d = 1$, this leads to maximizing b_{ii} over i, i.e., a D-optimal site has maximum prior variance. (Contrast this to the results above for A-optimality.)

E-optimality. E-optimality involves a search over K to maximize the minimum eigenvalue of

$$A^{-1}(K) = B^{-1} + K^T \Sigma^{-1} K.$$

For illustration, again assume $d = 1$. The incidence representation of a design $K(s)$ that observes at site s is $K(s) = (0, \ldots, 0, 1, 0, \ldots, 0)$, with 1 being in position s. Let b^{ij} (a^{ij}) denote the (i, j)th element of B^{-1} ($A^{-1}(s)$). Simple algebra implies that $a^{ij} = b^{ij}$ if $i \neq s, j \neq s$, and $a^{ss} = b^{ss} + \sigma_s^{-2}$. Thus, we can regard this problem as perturbing a selected diagonal element in B^{-1} by the corresponding precision (inverse of variance) of the measurement error, so that the minimum eigenvalue of the resulting perturbed matrix is largest.

Selection of a linear combination of X. We now relax the condition that K must be an incidence matrix, but assume $d = 1$ and the measurement error variance σ^2 is constant across sites. The observation is to be a *linear* combination of components of **X**. The problem is to find the optimal coefficients.

Let $U_1, U_2, \ldots, U_n$ and $\lambda_1 \geq \lambda_2 \geq \cdots \geq \lambda_n$ denote orthonormal eigenvectors (principal components) and the corresponding eigenvalues of B. Consider K of the form $K^T = \sum_{i=1}^n a_i U_i$ where $a_1^2 + a_2^2 + \cdots + a_n^2 = 1$. Note that $a_i = \langle K, U_i \rangle, i = 1, 2, \ldots, n$, and $\|K\|^2 = 1$. We can make this reduction with no loss in generality since the U_i span $\mathcal{R}^n$. Further, selecting K^T to be proportional to eigenvectors of matrices related to B has been suggested in the literature; see [Pal98].

With these definitions, we have that:

- A-optimality: maximize $\frac{KB^2K^T}{(\sigma^2 + KBK^T)}$. Hence, we are to maximize

$$\frac{\sum_{i=1}^n \lambda_i^2 a_i^2}{\sigma^2 + \sum_{i=1}^n \lambda_i a_i^2} \tag{5.16}$$

over the a_i, subject to $a_1^2 + \cdots + a_n^2 = 1$. The A-optimal solution is $a_1 = 1$, implying that $K^T = U_1$. To verify that

$$\frac{\sum_{i=1}^n \lambda_i^2 a_i^2}{\sigma^2 + \sum_{i=1}^n \lambda_i a_i^2} \leq \frac{\lambda_1^2}{\sigma^2 + \lambda_1},$$

perform the cross-multiplication and note that the conditions on the λ_i and a_i imply $\sum_{i=1}^n \lambda_i^2 a_i^2 \leq \lambda_1^2$ and $\sum_{i=1}^n \lambda_i^2 a_i^2 \leq \lambda_1 \sum_{i=1}^n \lambda_i a_i^2$.

- D-optimality: maximize $KBK^T = \sum_{i=1}^{n} a_i^2 \lambda_i$ over the a_i, $a_1^2 + \cdots + a_n^2 = 1$. It is easy to see that the optimal result is again $a_1 = 1$ and $K^T = U_1$.
- E-optimality: maximize the minimum eigenvalue of

$$A^{-1} = \sum_{i=1}^{n} \lambda_i^{-1} U_i U_i^T + \sigma^{-2} \sum_{i,j=1}^{n} a_i a_j U_i U_j^T$$

over the a_i, $a_1^2 + \ldots + a_n^2 = 1$. Since this is not a simple problem, it is natural to ask how well the A- and D-optimal solutions $a_1 = 1$ perform for the E-optimality criterion. Setting $a_1 = 1$, we obtain

$$A^{-1} = (\lambda_1^{-1} + \sigma^{-2}) U_1 U_1^T + \sum_{i=2}^{n} \lambda_i^{-1} U_i U_i^T.$$

Hence, the largest eigenvalue of A, when $K^T = U_1$, is $\max(\lambda_2, (\lambda_1^{-1} + \sigma^{-2})^{-1})$. Thus, if σ^2 is small relative to λ_1 (as it should be for the extra observation to be worth the trouble of designing and collecting), we would anticipate a reduction of worst-case predictive variance from λ_1 to λ_2.

The suggestion of observing along the leading eigenvector of B if K^T *is unrestricted* is consistent with the calculations of [Pal98] and [Ber99b]. However, in the context of adaptive observations for NWPs, this strategy does not take into account the practical constraints on admissible designs; it may be nearly feasible when U_1 happens to be highly localized in space. On the other hand, it may offer a novel strategy in smaller environmental monitoring problems.

2.3   Greedy Algorithms

Finding optimal designs when the number of observations d is small is feasible for spatial fields of moderate dimension (i.e., n is in the order of 100s). [Ber99b] considered optimal design in a toy weather model with 40 variables. Implementations typically took only a few minutes on a workstation.

However, the general problem of searching for optimal designs when both the number of observations d and the spatial dimension n are large is a challenging problem. Computer memory problems in storing and inverting large matrices arise. Exhaustive search over a very large number of designs may be infeasible. On the other hand, design constraints may reduce the number of feasible designs significantly. In practical adaptive observations problems, typical dimensions of $\mathbf{X}$ are on the order of 10,000,000. A complete search for the optimal K seems preposterous. However, there are typically additional conditions on the collection of admissible K. For example, the "home base" of the aircraft involved dictates limitations on the spatial range of the aircraft, yielding significant reductions in the cardinality of the set of admissible designs.

Still, there is a need to develop efficient search algorithms to reduce some of the computational burden. One possibility is to use sequential search algorithms, in which the d sites are chosen sequentially with just a few sites chosen at a time.

Consider taking a total of d observations "one-at-a-time," updating sequentially, and using the "current" posterior to design the next observation. Such a *myopic* algorithm can be very rapid, particularly in the context of D- (and A-)optimality. Assume all $\sigma_i^2 = \sigma^2$. As we have shown, the D-optimal single observation design is to observe a site, labeled s_l, for which the prior variance (diagonal elements of B) is maximized. Suppose we observe at s_l. As noted earlier, we can find the updated covariance matrix, A, using (5.5) before the observation is collected. This matrix now plays the role of "B" in determining the next site. A best single observation to take is a site having the largest variance for the new updated matrix. In terms of the original B, the result is a site i for which

$$b_{ii} - \frac{b_{is_l}^2}{\sigma^2 + b_{s_l s_l}}$$

is maximized. Proceeding sequentially until d observations are selected, this algorithm provides a comparatively rapid candidate design. It need not produce a D-optimal design for the simultaneous selection of d sites. However, its computational advantages in high dimensions make it a plausible approach.

Example Suppose $\sigma = 1$ and

$$B = \begin{pmatrix} 2.1 & 1.5 & 0.9 \\ 1.5 & 2 & 0 \\ 0.9 & 0 & 1.5 \end{pmatrix}.$$

The D-optimal solution is Sites 2 and 3, which, using (5.11), results in a posterior determinant of A equal to $1.305/7.5$. The sequential strategy first selects Site 1 and then Site 2. The implied determinant of A is $1.305/7.05$.

For comparison, the $d = 2$ A-optimal solution in this case is Sites 1 and 2; the sequential A-optimal algorithm also gives this design.

The greedy algorithms share some common features with variable selection in linear regression. This idea is suggested and used in [Nyc96]. In addition, combinatorial optimization methods, including stochastic search methods, may be useful.

3   Experimental Design in Space–Time

As in Section 1, let $\mathbf{X}_0, \mathbf{X}_1$, and $\mathbf{X}_2$ represent the state of the atmosphere at times, t_0, t_1, and t_2, respectively. The Bayesian adaptive observations problem can be stated as:

> At $t = t_0$, based on our prior distribution for $\mathbf{X}_1$, computed by combining our current distribution for $\mathbf{X}_0$ and the NWP dynamical

model, select additional observations, say $\mathbf{Y}$, to be taken at $t = t_1$ to optimize either the posterior covariance matrix of $\mathbf{X}_1$ or the posterior predictive covariance matrix of $\mathbf{X}_2$.

With some assumptions and simplifications, this idealized version of the adaptive observations in space–time can be reduced to a design problem as analyzed in Section 2. Suppose all available information at time t_0 about $\mathbf{X}_0$ is summarized through the distribution

$$\mathbf{X}_0 \sim N(\boldsymbol{\nu}_0, A_0). \tag{5.17}$$

Note that this distribution is itself a "posterior" in that it is based on all data available at time t_0. Alternatively, it is the result of data assimilation at t_0. In the parlance of data assimilation, $\boldsymbol{\nu}_0$ are "analyses"; A_0 is an "analysis error covariance matrix".

Next, assume that the NWP model uses dynamics given by

$$\mathbf{X}_1 = f(\mathbf{X}_0), \tag{5.18}$$

where f is a known, nonlinear function. The *tangent linear approximation* of this model is

$$\mathbf{X}_1 \approx f(\boldsymbol{\nu}_0) + F(\boldsymbol{\nu}_0)(\mathbf{X}_0 - \boldsymbol{\nu}_0), \tag{5.19}$$

where $F(\boldsymbol{\nu}_0)$ is the Jacobian of the transformation f, evaluated at $\boldsymbol{\nu}_0$. This leads to an approximate distribution for $\mathbf{X}_1$:

$$\mathbf{X}_1 \sim N(\boldsymbol{\mu}_1, B_1), \tag{5.20}$$

where

$$\boldsymbol{\mu}_1 = f(\boldsymbol{\nu}_0) \tag{5.21}$$

and

$$B_1 = F(\boldsymbol{\nu}_0) A_0 F(\boldsymbol{\nu}_0)^T. \tag{5.22}$$

Assume that the adaptive observations we are to collect at t_1 follow the model

$$\mathbf{Y} = K\mathbf{X}_1 + \boldsymbol{\varepsilon}, \tag{5.23}$$

where K is a $d \times n$ matrix and $\boldsymbol{\varepsilon} \sim N(0, \Sigma)$. Once collected, we combine these data with the *prior* for $\mathbf{X}_1$ given in (5.20), yielding the posterior

$$\mathbf{X}_1 | \mathbf{Y} = \mathbf{y} \sim N(\boldsymbol{\nu}_1, A_1), \tag{5.24}$$

where.

$$
\begin{aligned}
A_1 &= (B_1^{-1} + K^T \Sigma^{-1} K)^{-1} & \text{(5.25)} \\
&= B_1 - B_1 K^T (\Sigma + K B_1 K^T)^{-1} K B_1, & \text{(5.26)} \\
\boldsymbol{\nu}_1 &= A_1 (B_1^{-1} \boldsymbol{\mu}_1 + K^T \Sigma^{-1} \mathbf{y}) & \text{(5.27)} \\
&= \boldsymbol{\mu}_1 + B_1 K^T (\Sigma + K B_1 K^T)^{-1} (\mathbf{y} - K \boldsymbol{\mu}_1). & \text{(5.28)}
\end{aligned}
$$

In parallel to Remarks 1 and 2 of Section 2, these results are well known in NWPs, though alternate derivations are often used. First, an analog of kriging in space–time problems is the Kalman filter; more precisely, in our case, the *extended Kalman filter*, since linearization of the nonlinear dynamical model is used. Alternatively, penalized least squares derivations are the core of four-dimensional variational data assimilation. (See [Cou97], [Lor86], [Tar87].)

If our problem is to design for the best estimation of $\mathbf{X}_1$, we would optimize the covariance matrix A_1, as in Section 2. If our problem is to design for the best prediction of $\mathbf{X}_2$, the problem is more intricate. Suppose that

$$\mathbf{X}_2 = g(\mathbf{X}_1), \tag{5.29}$$

where g is a known, nonlinear function. (Since the time lags $t_1 - t_0$ and $t_2 - t_1$ may be different, we use g here rather than f.) Another tangent linear approximation on (5.29) implies that, *conditional* on the observations $\mathbf{y}$,

$$\mathbf{X}_2 | \mathbf{Y} = \mathbf{y} \sim N(\boldsymbol{\mu}_2, B_2), \tag{5.30}$$

where

$$\boldsymbol{\mu}_2 = g(\boldsymbol{\nu}_1) \tag{5.31}$$

and

$$B_2 = G(\boldsymbol{\nu}_1) A_1 G(\boldsymbol{\nu}_1)^T. \tag{5.32}$$

As before, G is the Jacobian of g evaluated at $\boldsymbol{\nu}_1$.

We note a major problem: G, hence B_2, are complicated functions of $\boldsymbol{\nu}_1$, which in turn, depend on data whose collection is being designed. In principle, the solution is easy: we should optimize the *expected* properties of B_2. In settings where the computation of such expectations is deemed infeasible, approximations may be needed. For example, one may replace $\boldsymbol{\nu}_1$ by its expectation $\boldsymbol{\mu}_1$. The resulting approximated B_2 is then optimized as in Section 2. These and other issues in adaptive observations for space–time processes are discussed more fully in [Ber99b].

4   Discussion

1. The evolution-of-state models in Section 3 essentially behaved as if the NWP model state was the "true" state in the sense that the NWP variable $\mathbf{X}$ served in the conditional distribution of real data. This is optimistic indeed. A potential relaxation of this assumption is to use a stochastic process model for $\mathbf{X}$, such as

$$\mathbf{X}_{t_{i+1}} = f_i(\mathbf{X}_{t_i}) + \eta_{t_{i+1}}, \tag{5.33}$$

where $\eta_{t_{i+1}}$ are random vectors intended to incorporate model error and external noise. If these error vectors are independent and follow $N(0, M_{t_{i+1}})$

distributions, then incorporating them in the design analysis is comparatively straightforward [Ber99b]. Also, see [Wik99c].

2. Section 3 considered an idealized version of adaptive observations. In general, a variety of *sequential* issues arise. First, various routine observations may become available before or after the adaptive observations will be collected, yet before the prediction time t_2. The analysis in Section 3 can readily be modified in such cases. However, the genuinely sequential problem of designing adaptive data collection over many epochs of collection and prediction times (and accounting for the cost of observation) is more difficult and is not addressed in the literature. See Chapter 7 of [Ber85] for an introduction to sequential decision analysis.

3. The notion of targeting special data collection to support physical model-based forecasting has a potentially wide application. Indeed, such thinking should not be confined to weather predictions. It may offer important strategies to a variety of environmental prediction problems.

4. Practical limitations in computing optimal designs were mentioned in Section 2.3. Making the optimizations discussed here operational is a research problem. An alternative promising idea is to consider optimal targeting on a relevant, but dimension-reduced, system.

5. It seems natural to jointly design future routine and adaptive observing strategies. That is, there are implications to the efficient design of fixed, observing networks and the design of routine satellite platforms, given that there is also an opportunity for adaptive observations. ([Ber99b] discuss developing adaptive procedures that account for routine observations, but not the reverse.) This suggestion extends well beyond the design for NWPs; it is relevant to the allocation of observational resources for analyzing arbitrary space–time processes.

6. Optimal procedures seem desirable, but mathematical optimization is based on assumptions and inputs. For us, adaptive observations designs were based on knowledge of the distribution of measurement errors, knowledge of the distribution of the "current state," and linearization of the dynamics. Robustness of results with respect to deviations from these inputs is an important issue. While the measurement error distribution can perhaps be reasonably specified, the latter two assumptions are not easily defended. Refinements of these assumptions are difficult and important. For example, the appropriate specifications of our matrix A_0 in (5.17) is problematic and is the subject of considerable research in data assimilation. Second, reliance on linearizations is also undesirable. The research area of *ensemble forecasting* is in part devoted to more sophisticated approaches to these issues. Interaction between such methods,x and the selection of adaptive observations, is of considerable interest [Bis99].

Seasonal Variation in Stratospheric Ozone Levels, a Functional Data Analysis Study

Wendy Meiring
University of California, Santa Barbara, CA 93106, USA

1 Introduction

Stratospheric ozone plays a vital role in restricting the ultraviolet radiation that reaches the surface of the Earth, as well as in controlling the atmospheric temperature distribution. The study of seasonal, chemical, and dynamical sources of variation in ozone levels is thus an important component of climate research [Wor95]. Stratospheric ozone depletion has been linked to anthropogenic pollutant emissions, including emissions of chlorofluorocarbons (e.g., [Bra99]). Additional chemical and dynamical forcings have been studied (e.g., [Sol96], [Ran96]). The detrimental impacts of stratospheric ozone depletion motivate the continued study of altitude-dependent chemical and dynamical sources of stratospheric ozone variability on a variety of time scales, building on a vast literature of published ozone trend studies (including [Tia90], [Log94], [Mil97a], [Wea98], and [SRN98]). We present functional data analysis methodology to study sources of variation in observations from balloon-based instruments, known as ozonesondes.

Ozonesondes measure the ozone concentration at many vertical levels as the balloon rises from the surface of the Earth and travels through the troposphere, lower and middle stratosphere (see [SRN98] for further details). Ozonesonde observations of the ozone vertical profile over a geographic location may be considered as a time series of samples from a vertical profile "curve" which evolves in time in response to dynamical and chemical sources of variability (see Figure 1). It is this concept of an irregularly sampled vertical curve, evolving in time, that leads us to consider functional data analysis methodology (e.g., [Ram97]). We use principal component analysis (e.g., [Jol86]), together with interpolating cubic splines (e.g., [Gre94]), to form a set of continuous functions such that each observed ozonesonde profile may be approximated by a linear combination of a subset of these continuous functions. The subset size and functions themselves are chosen so that a profile-specific linear combination will capture important features of the vertical structure of each of the ozonesonde profiles. Moreover, the time series of coefficients of each of these linear combinations (basis function approximations) will capture much of the dynamical and chemical variability in stratospheric ozone, provided that we include a "sufficient" number of basis functions in each linear combination. In this chapter we investigate seasonal dependence in the coefficients of these linear combinations, to study altitude-dependent seasonal variation in obser- vations from ozonesondes launched from Hohenpeissenberg in Germany. It is a natural extension to model altitude and seasonal-dependent trends as well as other dynamical and chemical sources of ozone variability, through modeling the coefficients of the basis functions in terms of these forcings. Extensions of this type provide another approach to investigate natural variability and an- thropogenic impacts on ozone levels. Further study of these complex processes is an essential component of climate change research.

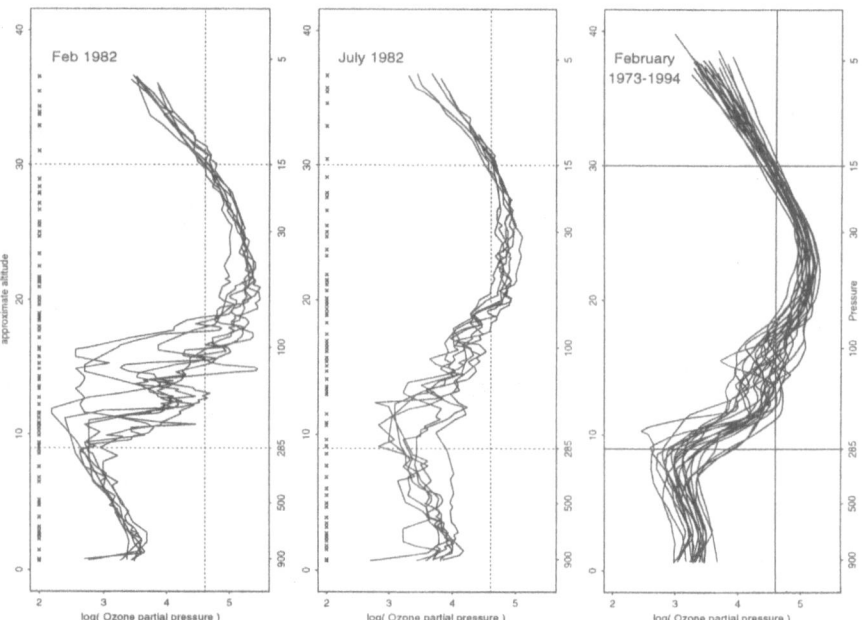

FIGURE 1. The first two panels show log(ozone partial pressure) for seven ozonesonde launches in each of February and July 1982. The left-hand axis of each plot shows the approximate altitude in kilometers, while the right-hand axis shows the atmospheric pressure in hPa. For one of the profiles in each of February and July 1982, the vertical levels at which observations are made are indicated by crosses at the left of each of the first two panels. The third panel shows lowess smooths [Cle79] through all February profiles separately for each year (each line corresponds to a smooth through all the profiles for February in 1 year). A vertical reference line has been drawn at log(100 micro-hPa).

2   Stratospheric Ozone Data

We obtained the Hohenpeissenberg ozonesonde records from the World Ozone Data Center, and analyzed observations from launches during the period 1973 through 1994. The first two panels of Figure 1 show ozone partial pressures[1] in micro-hPa on the log scale for seven ozonesonde launches in each of February and July 1982. This figure illustrates seasonal and interannual variability in the mean structure of the vertical profile, as well as seasonality and altitude dependence in the variability of stratospheric ozone levels. The left axis shows the approximate altitude in kilometers, and the right axis shows the atmospheric pressure in hPa. The ozonesonde measurements take place at irregular atmospheric pressure levels, that differ for each profile. The crosses at the left of each of the two panels indicate the vertical levels at which observations were made for one profile for each of the months. The third panel illustrates the interannual variability in ozone partial pressures by showing a lowess smooth[2] [Cle79] through all the February profiles separately for each year, giving one curve for each year in the period 1973 through 1994. In each of these panels, a vertical line has been drawn at log(100 micro-hPa) to aid comparisons.

We have used two transformations in presenting Figure 1: one, of the observed ozone partial pressures, and the other of the atmospheric pressure. We analyze ozone partial pressure on the log scale, since this transformation makes the observations more symmetric about their mean at each pressure level, and ensures that our estimates are positive. We present our analyses on a vertical scale of approximate altitude in kilometers, computed as a linear transformation of log(atmospheric pressure in hPa). A more accurate transformation between atmospheric pressure and altitude should depend on temperature (e.g., [Pei92]). The ozonesonde observations are more regularly spaced on the approximate altitude scale than on the original pressure scale.

Ozonesondes provide valuable information about the sources of variability in ozone partial pressures at different vertical levels, since they record ozone partial pressures at a higher vertical resolution than that available from current satellites. However, it is important to realize that each ozonesonde is carried by a balloon launched from a point in space and does not have the horizontal coverage of a satellite. Dynamical variation in circulation patterns may result in ozone being transported from a region with higher (or lower) ozone levels to the study area, since there are global patterns in the distribution of ozone (see, e.g., [Bra99]). This leads to substantial natural variability in stratospheric ozone levels over any one geographic location. Ozonesonde observations there-

---

[1] Partial pressure: the pressure which would be exerted by a particular component of a gaseous mixture if all other components were removed without other changes [McI91].

[2] For February of each year, and for each approximate altitude a, a weighted average is computed of all log ozone partial pressures observed at approximate altitudes within a local window of a. This provides a smooth curve through the February log ozone partial pressures as a function of altitude for each year. The weighting function is robust to outliers. The S-PLUS function *lowess* was used to compute each curve.

fore exhibit higher variability than spatial averages from satellite observations, and some smoothing may be necessary in investigating the sources of variability.

For simplicity we consider only those ozonesondes launched from Hohenpeissenberg that reach at least 15 hPa (approximately 30 km) prior to bursting. These ozonesondes were launched from Hohenpeissenberg at T irregularly spaced time points $\{t(j) : j \in 1, \ldots, T\}$. For each $j \in \{1, \ldots, T\}$, corresponding to time point $t(j)$, we have observations at $n(j)$ irregular heights in the vertical, namely at approximate altitudes $\left\{a_{1,j}, \ldots, a_{n(j),j}\right\}$, in decreasing order on the kilometer scale. We denote the log of the ozone partial pressure at height $a_{i,j}$ and time $t(j)$ by $z(a_{i,j}, j)$. We begin by interpolating the observations to n standard height levels, and denote these standard levels by $\{s_1, \ldots, s_n\}$ in decreasing order. We denote the $n \times T$ matrix of interpolated values by $\mathcal{X}$, and consider it to be T observations from a random vector X. The next step in the analysis of the profiles is to perform principal component analysis on this matrix. The background for this method is given in the next section.

3   Principal Component Analysis

Principal component analysis is an approach to finding the modes of variability that describe the majority of the variations in the observations. Consider the random vector $X = [X_1, \ldots, X_n]'$ representing true log ozone partial pressures at n vertical levels. Suppose X has finite population mean $E(X) = \mu = [\mu_1, \ldots, \mu_n]'$, and variance–covariance matrix

$$\Sigma = E\left[(X - \mu)(X - \mu)'\right],$$

where the (i,j)th element σ_{ij} is $\text{Cov}(X_i, X_j)$. If μ and Σ are known, the kth principal component of $X - \mu$ is the linear combination $w_k'(X - \mu) = \sum_{i=1}^{n} w_{ik}(X_i - \mu_i)$ where the $n \times 1$ vector of constants, w_k, is chosen to maximize

$$\text{Var}(w_k'(X - \mu))$$

subject to the constraints $w_k' w_k = 1$ and $\text{Corr}(w_k'(X - \mu), w_j'(X - \mu)) = 0$ for $j < k$. The kth population principal component loadings of $X - \mu$ are given by the kth column of the matrix $\Gamma = [w_1, w_2, \ldots, w_n]$.

In our study, we do not know the true distribution of the random vector X, and we estimate the population principal components by calculating sample principal components based on a "random" sample of T observations of the random vector X. (Our profiles are not a random sample, but principal component analysis still provides a means for us to estimate modes of variability.) Suppose initially we have observations at the same n vertical levels for all T time points, with $T > n$. (Irregular observations are first interpolated to n vertical levels, and will be discussed in Sections 3.1 and 4.) Suppose $\boldsymbol{x}_j = [x_{1j}, x_{2j}, \ldots, x_{nj}]'$ is the column vector of the jth observations of the random vector $\boldsymbol{X}$, and that the T observations are stored in an observation matrix

$$
\boldsymbol{\mathcal{X}} = \begin{bmatrix} x_{11} & x_{12} & \cdots & x_{1T} \\ x_{21} & x_{22} & \cdots & x_{2T} \\ & & \vdots & \\ x_{n1} & x_{n2} & \cdots & x_{nT} \end{bmatrix}. \quad \text{Let} \quad \hat{\boldsymbol{\mu}} = \begin{bmatrix} \frac{1}{T}\sum_{j=1}^{T} x_{1j} \\ \frac{1}{T}\sum_{j=1}^{T} x_{2j} \\ \vdots \\ \frac{1}{T}\sum_{j=1}^{T} x_{nj} \end{bmatrix}.
$$

In our application, $\boldsymbol{\mathcal{X}}$ is the matrix of the log ozone *values* (interpolated log ozone partial pressures) from T profiles at n vertical levels, with x_{ij} corresponding to the *value* (interpolated ozone partial pressure) on the log scale at level s_i at time $t(j)$. We now specify the principal components for the ozone profile values in terms of matrix decompositions. Let $\boldsymbol{\mathcal{Y}} = \boldsymbol{\mathcal{X}} - \hat{\boldsymbol{\mu}}\mathbf{1}'$ and so

$$
\hat{\Sigma} = \frac{1}{T}(\boldsymbol{\mathcal{X}} - \hat{\boldsymbol{\mu}}\mathbf{1}')(\boldsymbol{\mathcal{X}} - \hat{\boldsymbol{\mu}}\mathbf{1}')' = \frac{1}{T}\boldsymbol{\mathcal{Y}}\boldsymbol{\mathcal{Y}}'
$$

is an estimate of Σ. The singular value decomposition[3] of $\boldsymbol{\mathcal{Y}}'$ gives a $T \times n$ matrix U, an $n \times n$ diagonal matrix D, and an $n \times n$ matrix V such that

$$
\boldsymbol{\mathcal{Y}}' = UDV', \quad U'U = I_n, \quad V'V = I_n, \tag{6.1}
$$

where I_n is an $n \times n$ identity matrix. Now

$$
TV'\hat{\Sigma}V = V'VD'U'UDV'V = D^2,
$$

and the kth sample principal component of $\boldsymbol{\mathcal{Y}}$ is given by the kth column of $(U'\boldsymbol{\mathcal{Y}}')' = \boldsymbol{\mathcal{Y}}U = VD$. The kth column of U is known as the principal component loadings for the kth principal component. D^2 is also a diagonal matrix, with diagonal elements equal to the eigenvalues of $T\hat{\Sigma}$ in decreasing order of magnitude. An important statistic is

$$
\frac{\sum_{i=1}^{l} d_{ii}^2}{\sum_{i=1}^{n} d_{ii}^2}
$$

which represents the proportion of variance accounted for by the first l principal components. The first l principal components are the l uncorrelated simple linear combinations that explain the highest percentage of variation out of all sets of l uncorrelated simple linear combinations. For further details and proofs see, for example, [Mar79]. Other references include [Jol86] and [Pre88].

---

[3]Computed using the S-PLUS function *svd*.

3.1   Principal Component Analysis of the Interpolated Profiles

In order to calculate principal components using the methods described in the previous section, we need regularly spaced data to form a rectangular matrix such as $\mathcal{Y}$. Unfortunately, the ozonesonde observations are irregular, with measurements at different pressure levels for different launch days. Our approach has two steps:

1. estimate principal components based on linearly interpolating the observations to the n chosen levels for each launch date,
2. regress the *observed* log ozone partial pressures on continuous basis functions formed from these principal components, evaluated at the levels at which observations were made on this launch date.

In this section we illustrate the principal component analysis on n chosen levels for observations from Hohenpeissenberg. In Section 4 we present the regression on continuous basis functions.

We linearly interpolate each profile to $n = 22$ levels, uniformly spaced from approximately 9 km to 30 km, as illustrated in Figure 2 for one ozonesonde launch in February 1982, and another in February 1988. Methodology for choosing n remains an area of future research, and the choice of $n = 22$ is just a start. The squares in the right-hand panels of this figure show the interpolated ozone partial pressures on the log scale for each profile, corresponding to two columns of the matrix $\mathcal{X}$. The crosses in the right-hand panels show the means of the interpolated log ozone partial pressures calculated separately by vertical levels across all months and years, which is the centering vector $\hat{\mu}$ defined in the previous section.

For each $j \in \{1, \ldots, T\}$, the centered interpolated ozone partial pressures on the log scale for the ozonesonde launched at time $t(j)$ may be expressed as

$$x_j - \hat{\mu} \;=\; \sum_{k=1}^{n} u_{jk} P_k, \tag{6.2}$$

where P_k is the kth column of VD, and u_{jk} is the jth row and kth column element of matrix U in (6.1). This equation presents a complete decomposition of the centered interpolated log ozone values in terms of the principal components. If we include only the first $K < n$ terms in the summation of (6.2), we obtain an approximation of the interpolated values, plus a vector $e_j = [\, e_{1j} \ \ e_{2j} \ \ \cdots \ \ e_{nj} \,]'$ of residuals

$$x_j \;=\; \hat{\mu} + \sum_{k=1}^{K} u_{jk} P_k + e_j. \tag{6.3}$$

Figure 3 shows the contributions $\{u_{jk} P_k | k = 1, \ldots, 8\}$ in this approximation of the February 1982 (top row) and February 1988 (lower row) profiles of Figure 2. The left panel in each row shows the contributions of the first four

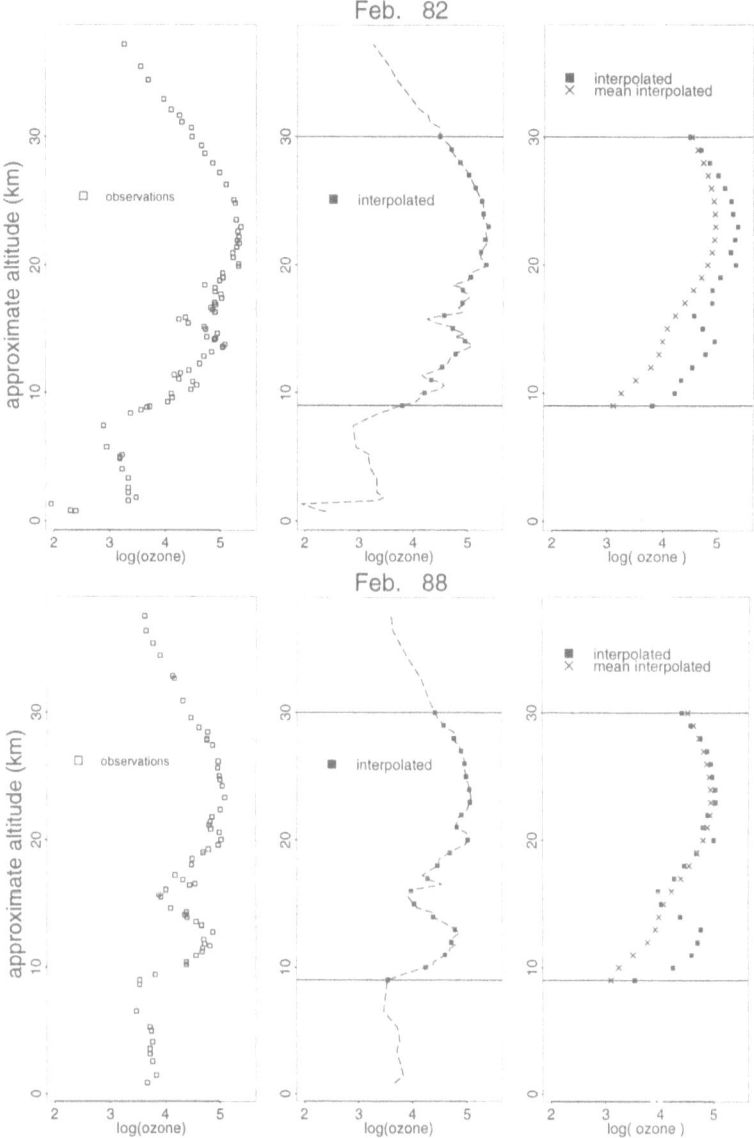

FIGURE 2. The top row of panels corresponds to one of the ozonesonde launches during February 1982. The second row of panels corresponds to one of the ozonesonde launches during February 1988. In each row, the left panel shows the ozonesonde observations on the log scale. The center panels, show the observations interpolated to $n = 22$ levels regularly spaced between 9 and 30 km in approximate altitude, using linear interpolation. The interpolated values are superimposed on a line joining the ozonesonde observations on the log scale, which were shown in the left panel. The right-hand panel shows the interpolated values, together with the mean of the interpolated values across all months and years, for each of the chosen 22 levels.

principal components; the contributions of the fifth to eighth are in the right panel. This illustrates that each principal component has a different relative importance in approximating the profiles for these two launch dates.

Figure 4 shows the interpolated values for each of the two launches (i.e., the columns of the matrix $\mathcal{X}$ corresponding to each of these two launch dates), together with the approximation in terms of the sum of the contributions of the mean interpolated profile and the first K principal components $\hat{\mu} + \sum_{k=1}^{K} u_{jk} P_k$ (6.3). These figures qualitatively show the approximation of each of these profiles by the subset of the first K principal components. In the left panel, $K = 4$. In the middle panel, $K = 8$. In the right panel, $K = 12$. As K increases, the standard deviation of the residuals decreases. As discussed in Section 3, the principal components are ordered in terms of the percent variation explained by each principal component. Figure 5 shows the cumulative percent variation explained by the first K principal components for the Hohenpeissenberg ozonesondes interpolated to $n = 22$ chosen levels, for $K \in \{1, \ldots, 22\}$. The total contribution of all 22 principal components gives a perfect fit to the centered interpolated values, since the principal components form an orthogonal basis for the 22-dimensional space. This "perfect fit" corresponds to the complete decomposition of (6.2). We wish to choose K to capture the main features of the profiles, but not to fit every detail on every profile since there is large natural variation on small time scales in the vertical. We are interested in variation on longer time scales.

Many approaches have been suggested for choosing K in principal component analysis. These include cross-validation [Wol78]. See, for example, [Jol86] for a review of techniques for choosing K in principal component analysis. Our aim is slightly different from that of regular principal component analysis though, in that we need to choose K sufficiently large so that we include sufficient basis functions to capture the main features in the profiles at all times of the year. In choosing K in this study we examined the fitted values obtained by regressing each profile on the continuous basis functions formed as cubic splines of the principal components (described in Section 4), and chose to use $K = 8$. Despite the large literature on this subject, a comprehensive statistical approach to choosing K remains as future research.

To illustrate the degree of fit of the interpolated values in terms of contributions of the first eight principal components, Figure 6 shows the variation in interpolated values which is unexplained by the first eight principal components. The first panel shows these residuals for all months taken together. The next four panels show the residuals for each of January, April, July, and October. While the median of the residuals is close to zero for each level when considering all months together, there is some evidence of bias at higher vertical levels for certain months. The variability in the residuals also appears to depend on season. This is due in part to different sources of variability in different seasons. An area of future research is how best to augment the set

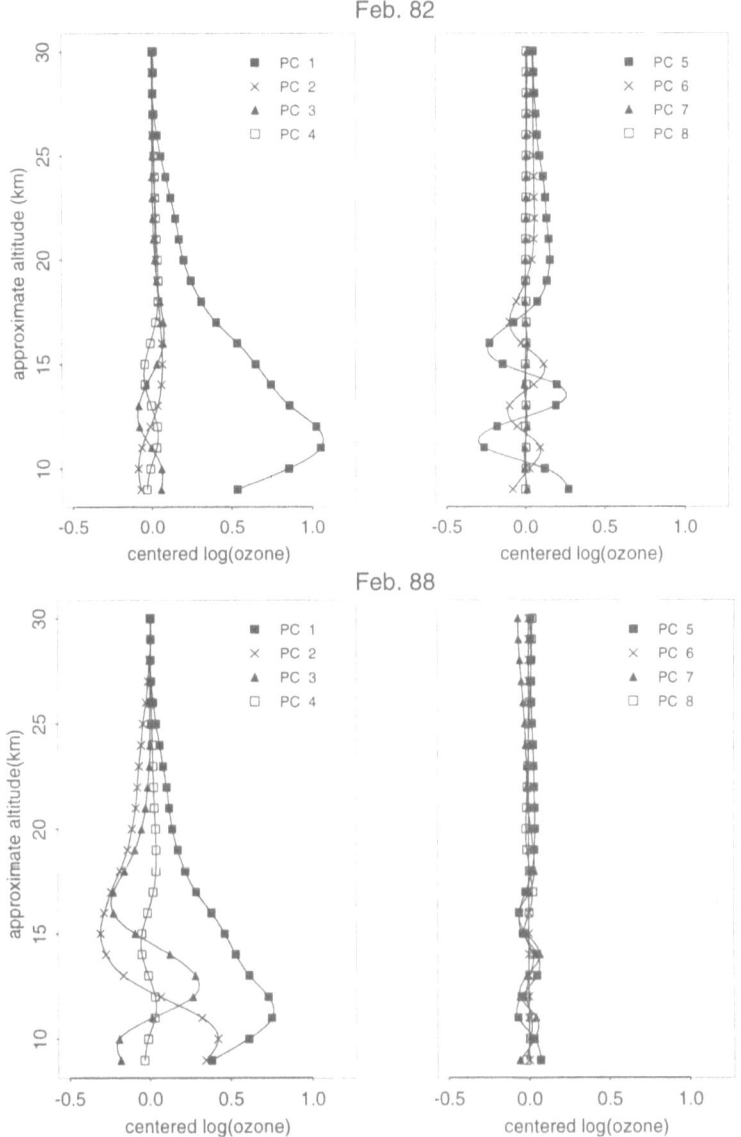

FIGURE 3. These panels illustrate the contributions from the first eight principal components in approximating the centered February 1982 profile and the centered February 1988 profile shown in Figure 2. Each profile first was centered by subtracting the mean interpolated profile displayed in the right-hand panel of Figure 2, followed by principal component analysis. The left panels show the contributions of the first four principal components; the right panel shows the contributions of the fifth to eighth principal components. These contributions are $\{u_{jk}\mathbf{P}_k | k = 1, \ldots, 8\}$ for launch time $t(j)$ in (6.3).

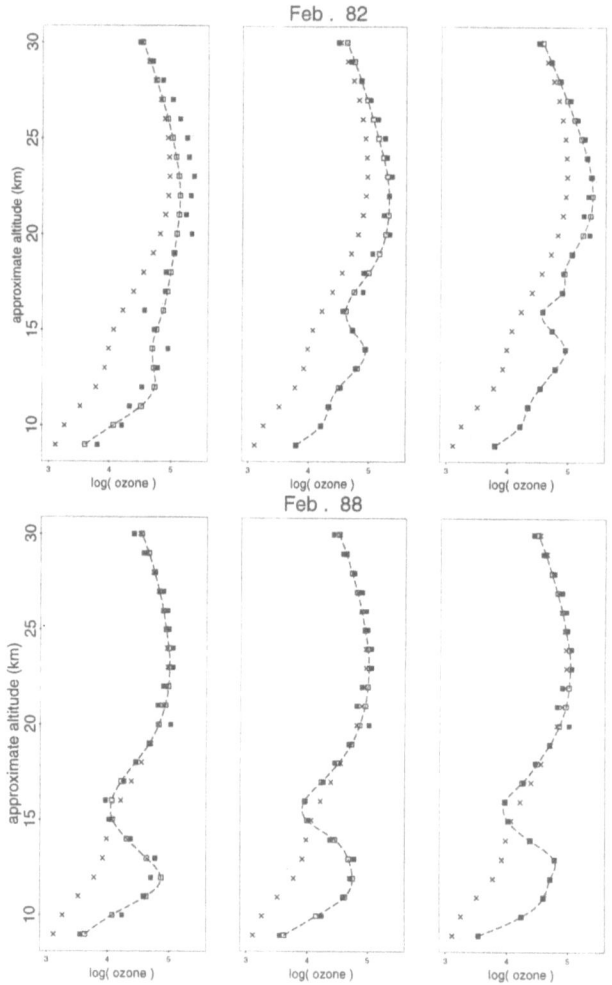

FIGURE 4. Each panel in the top row shows an approximation of the February 1982 profile shown in Figure 2, by the sum of the mean interpolated profile and contributions from the first K principal components. The lower row shows the analogous approximation of the February 1988 profile from Figure 2. For launch time, $t(j)$, the approximation is given by $\hat{\mu} + \sum_{k=1}^{K} u_{jk}\mathbf{P}_k$ (see (6.3)). In the left panel, $K = 4$. In the middle panel, $K = 8$. In the right panel, $K = 12$. In each panel, the approximation is shown by open squares joined by a curve. The solid squares show the interpolated observations, and the crosses show the mean interpolated profile, as in the right panel of Figure 2. The individual contributions of the first eight principal components were shown in Figure 3 for each of these profiles.

of basis functions to reduce these biases in separate months, and also how to take this into account in assessing the precision of estimates. The biases appear minimal (with the possible exception of October), at levels below 23 km, but adjusting for biases remains as future research.

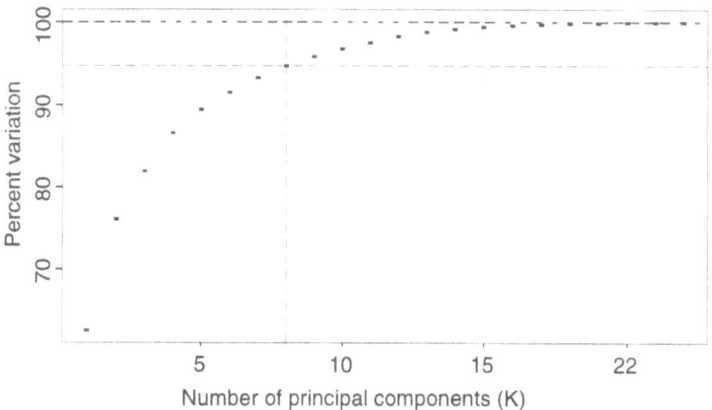

FIGURE 5. Percent variation of interpolated Hohenpeissenberg ozonesondes explained by the first K principal components, $K \in \{1, 2, \ldots, 22\}$.

4   Continuous Basis Functions to Represent Ozone

In the previous section we calculated the principal components of an $n \times T$ matrix of centered log ozone partial pressures, interpolated to n chosen vertical levels. The original observations are not on these chosen levels. As a second step in our analysis, we regress the centered log ozone partial pressures on continuous basis functions of altitude. These basis functions are simply the first K principal components interpolated by cubic splines.

Each of the 22 principal components may be considered in a natural way as a function of altitude. Let $p_k(s_i) = d_{kk}v_{ik}$ for $k \in \{1, \ldots, 22\}$ and $i \in \{1, \ldots, 22\}$, where s_i is the ith of the 22 chosen vertical levels considered in decreasing order and v_{ik} is the ith row and kth column element of matrix V in (6.1). Now

$$
P_k = \begin{bmatrix} d_{kk}v_{1k} \\ d_{kk}v_{2k} \\ \vdots \\ d_{kk}v_{nk} \end{bmatrix} = \begin{bmatrix} p_k(s_1) \\ p_k(s_2) \\ \vdots \\ p_k(s_n) \end{bmatrix}.
$$

For each $k \in \{1, \ldots, 22\}$ the cubic spline interpolant of the vectors P_k, as a function of altitude, is the curve $B_k(a)$ minimizing

$$
\int_{s_n}^{s_1} \left[\frac{d^2 B_k(a)}{da^2} \right]^2 da \quad \text{subject to} \quad B_k(s_i) = P_k(s_i) \quad \text{for all } i \quad \{1, \ldots, 22\}.
$$

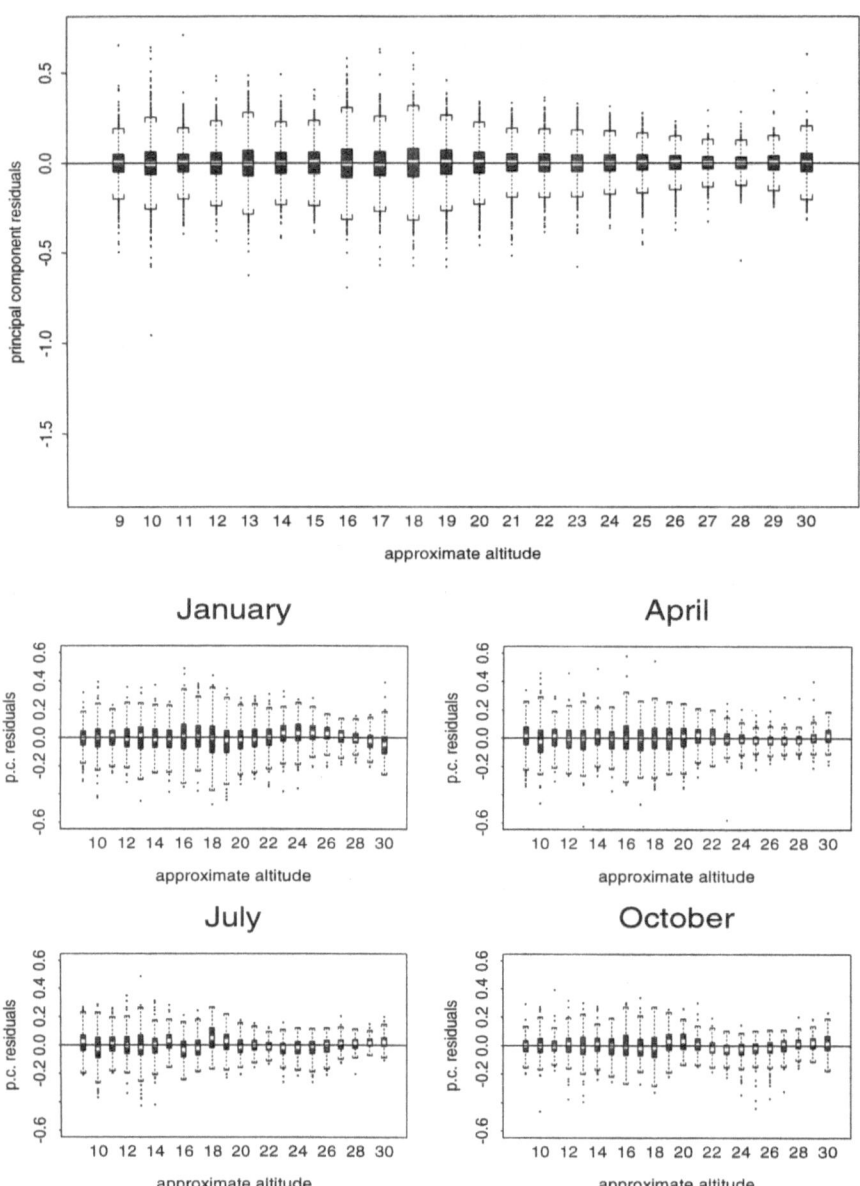

FIGURE 6. The first panel shows the residuals from an approximation of each interpolated profile of log ozone partial pressures in terms of the first eight principal components, plotted by vertical levels, for all months taken together. The second set of panels shows boxplots of these residuals by vertical levels plotted separately for each of four months, omitting residuals which are greater than 0.6 in absolute value.

The resulting curves $B_k(a)$ are unique, have continuous second derivatives for all $a \in [s_n, s_1]$, and are cubic polynomials in each of the intervals (s_{i+1}, s_i), $i \in 1, \ldots, n$ (e.g., [Gre94]).

Figure 7 shows the first eight basis functions $B_k(a)$, $k \in \{1, \ldots, 8\}$, formed as cubic spline interpolants of the principal components described in Section 3.1. In the first four panels, $B_k(a)$ is plotted, together with $B_k(a) + B_{k+1}(a)$ and $B_k(a) - B_{k+1}(a)$ for $k \in \{1, \ldots, 4\}$, respectively. These plots indicate how the basis functions modulate each other to explain local features of the vertical profile structure. In the last two panels, $B_k(a)$ are plotted for $k \in \{5, \ldots, 8\}$. We shall illustrate that these continuous functions capture much of the variation in ozone levels. However, this is only one of several approaches for forming continuous basis functions, some of which are discussed by [Ram97], including the smoothed principal components of [Sil96].

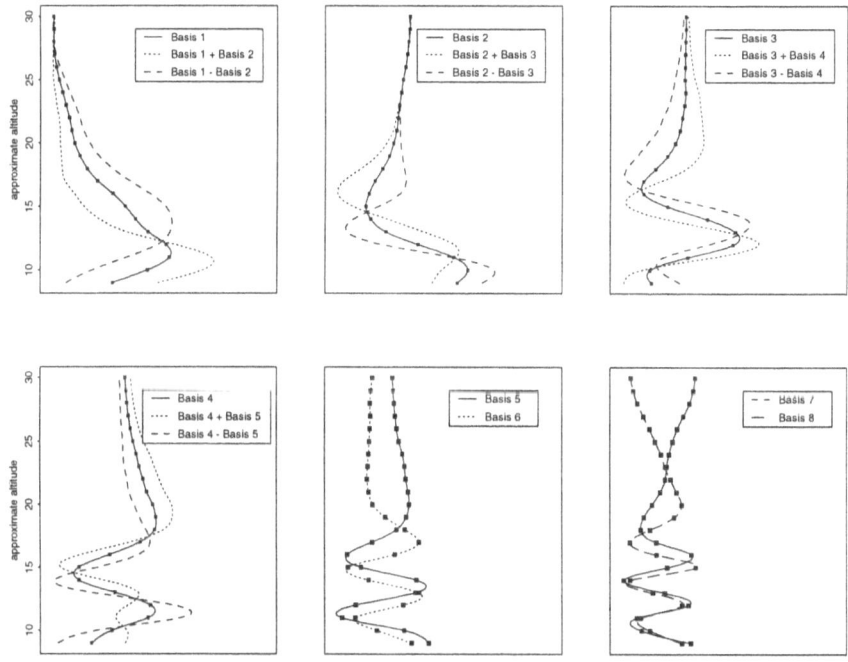

FIGURE 7. In each of the first four panels, the kth function $B_k(a)$ is plotted versus altitude as a solid line, together with $B_k(a) \pm B_{k+1}(a)$ for $k \in \{1, 2, 3, 4\}$, respectively. The next panel shows the 5th and 6th basis functions, followed by the 7th and 8th in the final panel. The dots on each on the kth basis function correspond to the kth principal component obtained using a singular value decomposition, $k \in \{1, \ldots, 8\}$. The vertical axis corresponds to approximate altitude in all panels.

4.1   Basis Function Coefficients

In the geosciences, data collected at irregular times and spatial locations are often interpolated to standard spatial locations and points in time. These interpolated data are then analyzed as if they are actual observations. The variability of the interpolated data is often substantially different from that of the original observations. While our basis functions are estimated from interpolated data, we use the original observations to estimate coefficients of the expansion of each profile with respect to the continuous basis functions. Specifically, for the jth profile $(t(j))$ we estimate coefficients $c_k(j)$ by regressing the observations on the first K basis functions, $B_k(a), k = 1, \ldots, K$, evaluated at the same vertical levels as the observations.

We now have a multivariate time series of coefficients

$$\{c_1(j), c_2(j), \ldots, c_K(j)\}, \qquad j \in 1, \ldots, T.$$

These coefficients model features in the curves and their evolution indicates a response to dynamical and chemical influences. Figure 8 shows scatter plots of $c_1(j)$ and $c_2(j)$ versus time $t(j)$. The means of the coefficients for each month of each year are included as a curve in each of the panels and vertical lines are drawn at January of each year. One of the most striking features is a seasonality in the first coefficient and seasonality is also present in the time series $c_2(j)$. There is also variability on other time scales, both longer and shorter than the seasonal cycle.

5   Varying Coefficient Models

The statistical challenge now is to understand the factors influencing the profile shape through modeling the coefficients. For illustration, we first consider the seasonal cycle. In this case we expand the coefficients in terms of sines and cosines:

$$c_k(t) \;\;=\;\; \nu_k + \sum_{q=1}^{Q} \left(\alpha_{qk} \cos \left(\frac{2\pi qt}{365} \right) + \beta_{qk} \sin \left(\frac{2\pi qt}{365} \right) \right) + \eta_k(t).$$

Here t is the julian date of the launch and ν_k, α_{qk}, β_{qk} are estimated by least squares. We denote the fitted coefficients by $\hat{c}_k(t)$. Figure 9 shows the fitted seasonal cycles for the first, second, and seventh coefficients in terms of $Q = 2$ Fourier components. The seasonal component accounts for different proportions of variability for the different coefficients. The first, second, and seventh are displayed to illustrate differing seasonal cycle patterns in different coefficients, accounting for different modes of variability. The seasonal cycle accounts for a substantial portion of variability in these three coefficients.

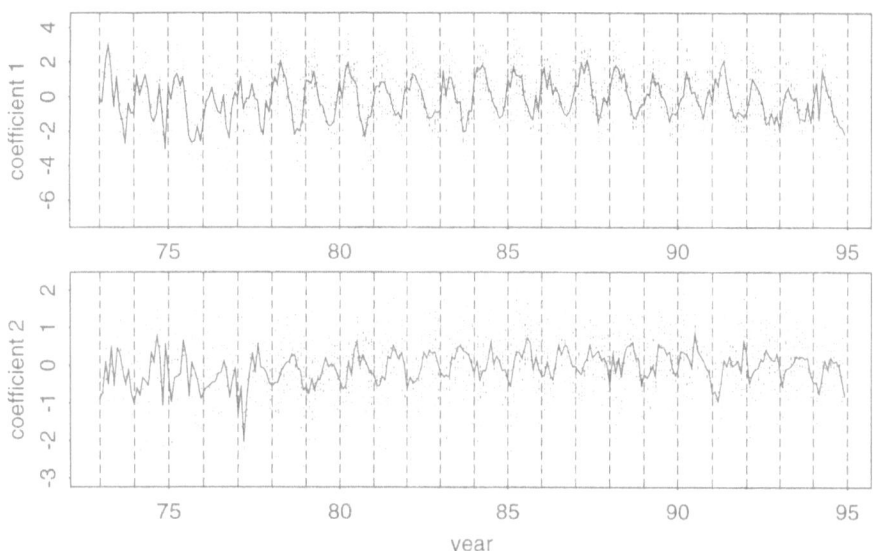

FIGURE 8. Scatter plots of the first and second coefficients of the basis function expansion versus time. A curve joins the means of all profiles for each month in each year. Vertical lines correspond to January of each year.

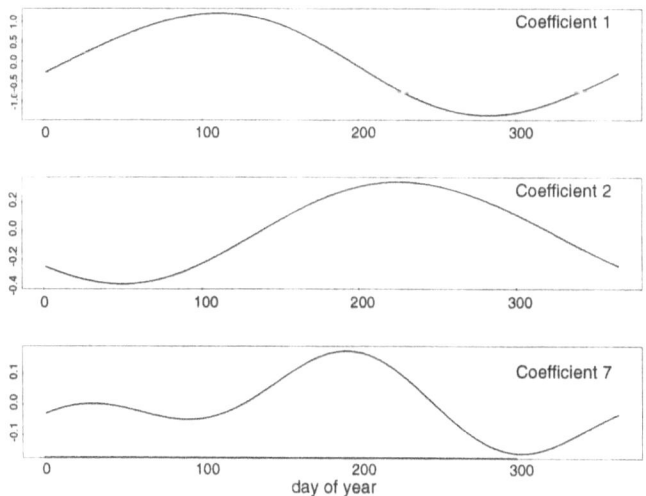

FIGURE 9. The fitted seasonal cycles for the first, second, and seventh coefficients in terms of $Q = 2$ Fourier components.

The fitted log ozone value $\hat{Z}(a, d)$ for approximate altitude a and day of year d is given by

$$\hat{Z}(a, d) = \sum_{k=1}^{K} \hat{c}_k(d) B_k(a).$$

The top panel of Figure 10 shows $\hat{Z}(a, d)$ for $a \in \{10, 14, 18, 22, 26, 30\}$ on the approximate altitude scale. The lower panel is a contour plot of the fitted log ozone values for different days of the year and approximate altitudes.

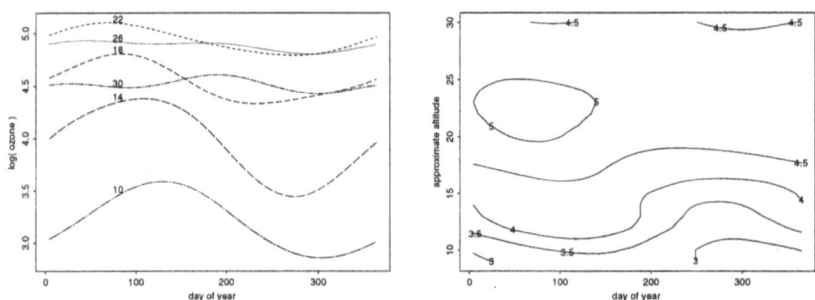

FIGURE 10. The fitted seasonal cycle for different vertical levels. Top panel: plots of the fitted log ozone values for approximate altitudes of 10, 14, 18, 22, 26, and 30 kilometers. Lower panel: contour plot of fitted log ozone values versus approximate altitude and day of year.

The seasonal cycle accounts for substantial variability in the stratospheric ozone levels. However, there are many processes acting on different spatial and temporal scales. The seasonal cycle alone is insufficient to account for nonlinear trends and interannual variability in the coefficients (and therefore also insufficient to model variability in the vertical stratospheric ozone profiles). Figure 11 shows the coefficients $c_1(t)$ for all January ozonesonde launches plotted versus year, showing a nonlinear trend which may be due to circulation or chemical factors. A cubic smoothing spline (e.g., [Gre94]) is superimposed on these coefficients to highlight the trend. Sources of variability, on different spatial and temporal scales, can be investigated through a model such as

$$Z(a, t) = \sum_{k=1}^{K} \tilde{c}_k(t, W) B_k(a) + \sum_{k=1}^{K} \eta_k(t) B_k(a) + \epsilon(a, t),$$

where W is a collection of time series of dynamical and chemical measures and

$$c_k(t) = \tilde{c}_k(t, W) + \eta_k(t)$$

is a model of the coefficients in terms of these dynamical and chemical components of variability, including nonlinear trends.

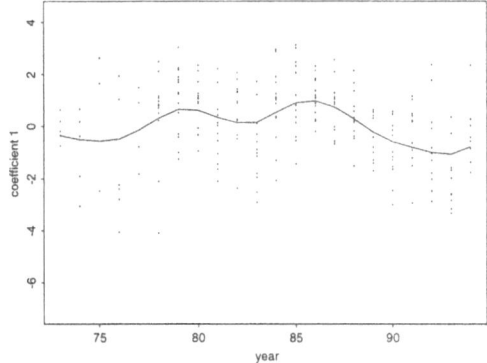

FIGURE 11. The coefficients $c_1(t)$ for all January ozonesonde launches, plotted versus year. The solid line is a cubic smoothing spline through these points (e.g., [Gre94]).

6   Discussion

The key idea illustrated by this ozonesonde study is that an analysis of profile shape can be reduced to modeling a small number of coefficients. As an example we have modeled the altitude-dependent seasonal cycle in ozone partial pressures above Hohenpeissenberg. The models presented in this chapter are being extended to detect interactive sources of variability on different time scales at different vertical levels. [Hol95] and [Ran96] have noted variation in satellite measured ozone levels on the time scale of the dynamical quasi-biennial oscillation in tropical wind directions (e.g., [Pei92]). [Sol96] and [Sol98] have studied chemical processes associated with aerosols after major volcanic eruptions such as Mount Pinatubo, and mid-latitude ozone depletion after Mount Pinatubo has been indicated in profile analyses (e.g., [Hof94]). We are investigating whether modeling the basis function coefficients will enable us to examine further processes on these dynamical and chemical time scales. Computational techniques for a proposed modeling approach are presented in [Mei98a], made possible by the dimension reduction due to basis function expansion as described in this chapter.

Acknowledgments

The author is grateful to Doug Nychka for numerous helpful discussions and collaboration on functional data analysis, and to both Mark Berliner and Doug Nychka for research encouragement and helpful comments on a draft of this chapter. The author has benefitted from numerous discussions with scientists at the National Center for Atmospheric Research, although any errors in interpretation are her own. This research was primarily funded by the National Center for Atmospheric Research Geophysical Statistics Project, sponsored by the National Science Foundation under Grant #DMS93-12686; and by a Regents Junior Faculty Fellowship from the University of California, Santa Barbara. The author is grateful to the World Ozone Data Center for providing the ozonesonde data.

Neural Networks: Cloud Parameterizations

Barbara A. Bailey
University of Illinois at Urbana-Champaign, IL 61820, USA

L. Mark Berliner
Ohio State University, Columbus, OH 43210, USA

William Collins
National Center for Atmospheric Research, Boulder CO 80307, USA

Douglas W. Nychka
National Center for Atmospheric Research, Boulder, CO 80307, USA

Jeffrey T. Kiehl
National Center for Atmospheric Research, Boulder, CO 80307, USA

1 Introduction

The role of clouds in the climate system is very complex and is the subject of much interest and research. Clouds interact nonlinearly with radiative, dynamical, chemical, and hydrological processes in the atmosphere on a wide range of temporal and spatial scales. Clouds play a fundamental role in controlling the amount of solar and infrared radiation available to the climate system. The radiative properties of clouds make them a key component in the energy balance of the Earth. In particular, clouds are involved in both heating and cooling in the determination of the Earth's temperature. On average, roughly 50% of the Earth is covered by clouds. They contribute to the planet's *albedo* by reflecting *some* incident sunlight (shortwave radiation) back to space (they also absorb some). However, they also partially block the escape of infrared radiation from below; that is, they exert a *greenhouse effect* on Earth. (Clouds are the primary contributors to the greenhouse effect.) They also emit some longwave radiation. Clouds also play an essential role in controlling the amount of moisture available to the climate system. Through precipitation, clouds serve as a conduit for the transfer of heat from the oceans to the atmosphere. They are also important in many chemical processes such as the absorption of water-soluble chemicals and pollutants in cloud droplets and their elimination by precipitation. See [Tre92] for further discussion.

Atmospheric general circulation models (GCM) are global numerical models that predict the time evolution of atmospheric variables including temperature, humidity, and pressure in three dimensions in space. These models comprise complex mathematical equations that describe these physical processes and their interrelationships. The equations are too complex to be solved exactly. Instead, versions of the mathematical equations are discretized in both time and space, encoded numerically, and run to produce approximate solutions. In this process, the Earth's atmosphere is divided into a discrete number of grid boxes. Even the most complex GCM is limited in the spatial detail it can *resolve*. Typical GCMs have a horizontal grid of 2.5° × 2.5° (250 × 250 km at the equator) and on the order of 20 vertical layers.

Many important climate processes occur over scales smaller than an individual GCM grid box. In particular, most clouds are smaller in area than the typical grid resolution of a GCM. This is our puzzle: on one hand, clouds are critical to the overall behavior of the climate; on the other hand, practical limitations currently require GCMs to have resolutions too coarse to model clouds directly. Approaches to dealing variables acting at scales below the resolution of the model are known as *parameterizations*, or, more precisely in our case, *subgrid-scale parameterizations*. Through parameterization, the amount of cloud cover (or its fraction) at the model grid box level is introduced into a GCM.

Simply put, clouds are visible structures of condensed water vapor. Most water vapor is the result of evaporation and *convection* over the oceans. A process involving vertical transport caused by buoyant instabilities (e.g., water vapor rising) is known as convection. In current GCMs, to estimate upper level cloud amounts, a gridbox is determined to be in a "convective" or "nonconvective" state by using the vertical layer profile of several indicator variables over the gridbox. The cloud amount is then calculated using a relatively simple function (linear or quadratic) of relative humidity [Tre92]. Most cloud amount prediction schemes are too simplistic to accurately model the real spatio–temporal distribution of cloud cover. Because of the complex interactions of cloud occurrence, their altitude and water content, and the microphysical properties of condensed water that influence cloud radiative properties, clouds remain one of the principal unresolved aspects of climate modeling [Tre92].

One requirement of a cloud parameterization is that it predicts the amount of cloud cover on a gridbox at time t based on model variables or functions of model variables that are resolved by the model and computed at the previous time step. Our modeling objective is to develop a statistical model for the subgrid-scale spatial and temporal distribution of cloud cover by linking large scale climate model variables, such as relative humidity, with cloud cover. By viewing the parameterization problem as one of building a statistical model, we offer the opportunity not merely to construct a parameterization formula, but also to provide quantifications of the uncertainties associated with that parameterization in both space and time.

To help set the stage for relating statistical models to parameterizations, we offer a mathematical explanation of parameterizations in Section 1.1. Section 1.2 is a general description of neural network models and their uses in estimating complex statistical relationships among variables. Versions of neural network statistical cloud parameterizations are formulated, fit to data, and analyzed in Section 2. In Section 3 we use models developed in Section 2 to simulate cloud cover as a test of their value. Discussion and further comments are given in Section 4.

1.1 The Essence of Parameterization

Suppose a dynamical system model for selected atmospheric variables is represented as follows (see [Ber99c]): First, the variables are separated into two groups, say x and y. The complete dynamical system is given by

$$x_{t+1} = h(x_t, y_{t+1}),   \tag{7.1}$$

$$y_{t+1} = g(x_t, y_t).   \tag{7.2}$$

The complete system is too large and complex for practical implementation. The x variables are considered crucial for modeling or at least selected for modeling, subject to practical limitations. Hence, we build a dynamical system for x only.

The basic steps are:

1. Choose a summarizing function (or functions), $C(y)$, which serves as a dimension reduction. Namely, it summarizes the high-dimensional vector y via a relatively low-dimensional summary, C. Associated with the dimension reduction, suppose there is a companion dynamical function, say $\tilde{h}$, such that (7.1) is reasonably approximated by

$$x_{t+1} = \tilde{h}(x_t, C_{t+1}).   \tag{7.3}$$

2. Construct a *parameterization*:

$$C_{t+1} \approx f(x_t, \theta),   \tag{7.4}$$

   where θ is some collection of parameters.

3. Substitute (7.4) in (7.3), leading to an approximating system:

$$x_{t+1} = \tilde{h}(x_t, f(x_t, \theta)).   \tag{7.5}$$

The key is that (7.5), our GCM, only involves x. Relating back to our context, x represents the collection of physical variables that are to be resolved by the GCM, while y represents various unresolved physical quantities. The selection of C in our case is that we model cloud cover in two dimensions, rather than individual clouds.

The relationship in (7.4) is an approximation. The development of this approximation will be viewed as a statistical regression problem. Included in the development are

1. the selection of variables or functions of x to actually be used in the model;
2. the estimation of model parameters θ; and
3. the associated uncertainties.

That is, the statistician replaces the approximation symbol in (7.4) by considering a stochastic model

$$C_{t+1} = f(x_t, \theta) + \text{error}, \qquad (7.6)$$

and suggests a statistical model for the errors. This leads to quantification of the approximation. Finally, note that this argument demonstrates a role for statistical thinking even against a background of deterministic modeling. While h and g might be thought of as deterministic, uncertainty about them and/or inability to compute them leads to a statistical component in the GCM.

An important enhancement to the above simple explanation is used in this chapter. Inspection of (7.2) suggests an alternative parameterization in the spirit of time series modeling. Namely, we generalize (7.4) to

$$C_{t+1} \approx f(x_t, C_t, \theta). \qquad (7.7)$$

Since C_t was computed in the previous time step, this is a legitimate parameterization. Converting to the statistical viewpoint, our modeling task is to build a regression time series model

$$C_{t+1} = f(x_t, C_t, \theta) + \text{error}. \qquad (7.8)$$

1.2   Neural Networks and Fitting Nonlinear Models

From the statistical viewpoint, we are to construct a regression model describing the space–time evolution of a complex process (cloud cover). A first-order time-lagged, spatial nearest-neighbor model is used. By "first-order time-lagged," we mean that only information at time t is used to predict the cloud cover amount at time $(t+1)$. By "spatial nearest-neighbor," we mean that predictions of cloud cover in a particular gridbox (or site) at time $t+1$ are based on information available at time t at that site as well as nearby sites. In two dimensions, consider a regular (equal area boxes), gridded region with r rows and c columns. Define $L = rc$. Let $\boldsymbol{X}_t = (x_{1t}, \ldots, x_{Lt})'$ and $\boldsymbol{Y}_t = (y_{1t}, \ldots, y_{Lt})'$ be an L-vector of values at time t. (Each of the x values may themselves be vectors but this is not reflected in the notation.) A first-order lagged nearest-neighbor statistical prediction model for each grid site l is

$$Y_{l,t+1} = f(\boldsymbol{X}_{\mathcal{N}(l),t}, \theta) + e_{l,t+1}, \tag{7.9}$$

where $e_{l,t}$ represent random model errors (here we assume them to be mutually independent, Gaussian random variables with mean 0 and common (unknown) variance σ^2 across both space and time). In general, $\mathcal{N}(l)$ represents a *neighborhood*; a collection of sites defined to be neighbors of site l. In this chapter we choose $\mathcal{N}(l)$ to be the site l itself and its four nearest-neighbors; that is, the nearest-neighborhood for site l, with row–column index (i, j), is the set

$$\mathcal{N}(l) = \{(i, j), ((i + 1), j), ((i - 1), j), (i, (j + 1)), (i, (j - 1))\}.$$

The nearest-neighborhood structure induces spatial dependence among sites. Note that there is a problem near the boundaries of the region: some sites do not have modeled neighbors. For the purposes of this chapter, we take the most expedient solution. We only fit the model for sites with complete neighbor information. For our region, we only fit the model at $L^* = (r-2) \times (c-2)$ sites. Our goal is to develop a test-of-concept, rather than a final parameterization. To that end, we build the model in a limited spatial domain (described in the next section). While extending the model to the entire planet offers many challenges, clearly there would be no boundary problems.

The map f in (7.9) is estimated by a feed-forward *neural network* with a single layer of hidden units [Ell92a]. A neural network is a very flexible nonlinear model. Neural networks are used because the relationships among the various variables modeled here, though thought to be extremely complicated, are not explicitly known. For each time t, the form of the map is

$$f(\boldsymbol{X}) = \beta_0 + \sum_{i=1}^{k} \beta_i \varphi(\boldsymbol{X}^T \boldsymbol{\gamma}_i + \mu_i), \tag{7.10}$$

where $\varphi(u) = e^u / (1 + e^u)$ and the parameter vectors $\boldsymbol{\beta}$, $\boldsymbol{\gamma}$, and $\boldsymbol{\mu}$ are unknown. If $\boldsymbol{X}$ is a d-dimensional vector, $\boldsymbol{\beta}$ has length $k + 1$, $\boldsymbol{\mu}$ has length k, and the $\boldsymbol{\gamma}_i$ are k vectors each of length d. The total number of parameters in the model is equal to $1 + k * (d + 2)$. As is common, these parameters are estimated by nonlinear least squares.

The *complexity* of the model, traditionally defined to be the number of hidden units k in the model, is chosen based on generalized cross-validation (GCV). Cross-validation is a standard approach for selecting smoothing parameters in nonparametric regression [Wah90]. The GCV function is defined as

$$\text{GCV}(p) = \frac{\frac{1}{n}\text{RSS}_p}{\left(1 - p\frac{c}{n}\right)^2},$$

where n is the number of data points used to fit the model; p is the number of parameters in the model; and RSS_p is the residual sum of squares (i.e., the sum of squared differences of data values and corresponding fitted values) for the best-fitting p-parameter model. Parameters enter the neural network model of (7.10) in groups as the number of hidden units increase. The model-fitting strategy is to fit models with a range of hidden units. The model order is then chosen to be that p for which the decay in $GCV(p)$ "levels off." The idea is that the numerator of $GCV(p)$ would be decreasing in p; the denominator is intended to appropriately moderate that decay. The cost parameter, c, is an ad hoc adjustment to increase $GCV(p)$ for larger values of p. The standard GCV function is $GCV(p)$ with $c = 1$, but we prefer cost values $c = 2$ to guard against overfitting.

In most cases, the nonlinear least squares objective function for fitting a neural network has a very large number of local minima. A preliminary stochastic search is therefore often suggested to obtain an approximate solution. The following short description illustrates the minimization procedure and the care taken to find the least squares estimate. From a very large collection (250,000) of starting points, i.e., specifications of the parameters, a specified number (250) of parameter points with the lowest root-mean-square error (RMSE) are used as initial points in an iterative procedure for minimization of the RMSE with a moderate convergence tolerance. A specified number (20) of these "solutions" are then used as starting points in the minimization procedure, but with a smaller convergence tolerance. The parameter set with the smallest RMSE is taken to be the least squares estimate [Nyc]. The above procedure is repeated for each value of k in the range of interest.

2   Cloud Parameterizations

The primary dataset available for building the model is three months of hourly infrared radiation (IR) temperature satellite data collected during the TOGA COARE (Tropical Ocean Global Atmosphere Coupled Ocean- Atmosphere Response Experiment) [Vel94]. Figure 1 is an example of an IR satellite image for a particular hour on February 9, 1994. From the IR temperature data, individual clouds have been identified by a numerical algorithm that starts from the coldest pixel center of the cloud and then searches outward from the center. It stops at a specified threshold temperature and the pixels for that individual cloud are identified. Every pixel in the image is identified as being a cloud or no cloud. The cloud cover is the 0–1 (no cloud/cloud) pixel count of the image. The resolution of the cloud count data is a 0.045° or 5 km square grid.

To develop parameterizations for a GCM at the 2.5° resolution described in Section 2.1, our cloud data are obtained by summing the original cloud counts over the larger gridbox. There are approximately 58^2 cloud count pixels for

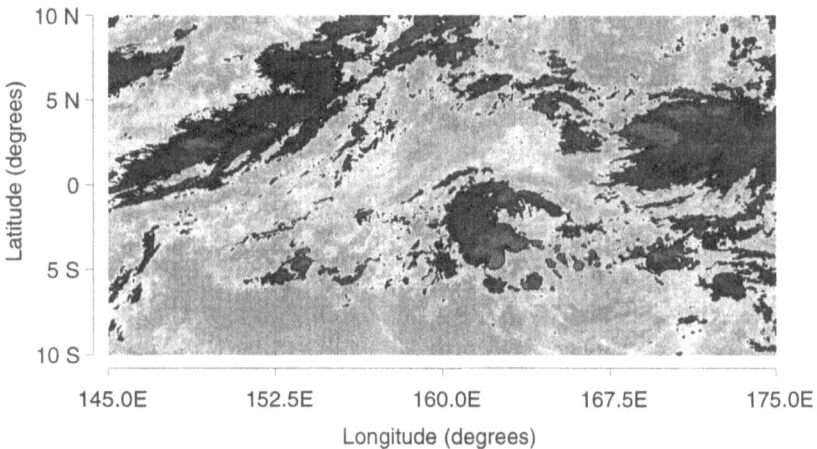

FIGURE 1. An hourly IR satellite image over the TOGA COARE region.

each $2.5°$ gridbox. Let $C_{l,t}$ be the resulting cloud cover count at the $2.5°$ grid level for site l at time t. Finally, note that the $2.5°$ gridding over the TOGA COARE region leads to $L = 8 \times 15$ sites. Recall that eliminating boundaries means that we actually fit $L^* = 6 \times 13$ sites at each time.

2.1   A Preliminary Model

As a preliminary test for the statistical prediction of cloud cover, we constructed a model using *outgoing longwave radiation* (OLR) measured in (Watts/meters^2) as the predictor or explanatory variable. Roughly 37% of the planet's OLR is attributable to clouds (the balance due primarily to emissions from atmospheric component gases (H_2O, CO_2, etc.) as well as a small fraction as emission from the surface). However, clouds absorb and return more longwave radiation than they emit. Hence, OLR is comparatively small for cloudy regions and large for clear sky regions.

Note that there is strong evidence for a complex, nonlinear relationship between OLR and IR data. The following exercise tests whether a nearest-neighbor neural network can approximate this nonlinear relationship.

We developed an OLR dataset at the $2.5°$ grid resolution for the TOGA COARE region and period. Hourly OLR data were obtained using aircraft-borne instruments calibrated to measure radiation in a selected frequency band during the TOGA COARE. For our purposes we define $OLR_{l,t}$ to be the value of OLR at grid site l at time t for use in the regression model. These values were all computed as averages of the aircraft OLR data over the appropriate $2.5°$ gridboxes.

A first-order time-lagged, spatial nearest-neighbor model for $C_{l,t}$, based on OLR, is

$$C_{l,t+1} = f(\mathbf{OLR}_{\mathcal{N}(l),t}, \theta) + e_{l,t+1}. \tag{7.11}$$

The model in (7.11) was fitted to hourly data for 36 hours starting 0Z on February 9, 1993. Figure 2 summarizes those results based on eight hidden units. Figure 2(a) is an image plot of residuals. Time in hours is displayed on the horizontal axis; the $L^* = 78$ sites are labeled vertically. The 78 sites are numbered so that first eight sites are the first column of the gridded region, sites 9–16 are the second column of the gridded region, etc.

There does not seem to be a systematic pattern in the residuals. In Figure 2(b) RMSE (summed over sites) is plotted over time. Note the fluctuations of about 7–20% RMSE with an average RMSE of about 13%. Percents are the fraction of the total number of pixels in a gridbox (58^2), multiplied by 100. Mean relative error at time t is defined to be

$$\frac{1}{78} \sum_{l=1}^{78} (C_{l,t} - \hat{C}_{l,t})/\hat{C}_{l,t}, \tag{7.12}$$

where $\hat{C}_{l,t}$ is the predicted cloud cover amount. Figure 2(c) is a plot of the mean relative error at each site over time; results range from -0.29 to 0.35.

An image plot of site-wise RMSE (averaged over time) is shown in Figure 3(a). To examine more closely the model fit at individual sites, four sites, with varying RMSE, were examined by plotting the observations and the fitted values over time. The locations of the sites are indicated in Figures 3(a). Figures 3(b)–(e) are time series plots for four sites. The solid lines represent data values; dotted lines indicate fitted values. These plots suggest that the model is capturing cloud cover evolution at these sites.

To examine out-of-sample prediction error, the fitted model was used to predict cloud cover for the next 28 one-hour time steps, based on OLR data for that time period. Figure 4 is a summary of the predicted cloud cover for the 28 hours. (In these figures, and for the balance of the chapter, we use the root-mean-squared *prediction* error (RMSPE) to emphasize that models were not refit using data in this period.) The magnitudes of errors in Figures 4(b) and (c) are very similar to the errors in Figures 2(b) and (c), indicating that the model is stable and predicts reasonable cloud cover amounts.

These results lend hope that the use of a first-order lagged nearest-neighbor neural network model may be productive in predicting cloud cover changes over time and space.

2.2   Toward a GCM Cloud Parameterization

We now consider the development of a first-order lagged nearest-neighbor neural network model in the spirit of a parameterization. We use four classes of predictor variables. First, following the general formulation of parameterization reflected in (7.8), we use $C_{l,t}$'s in predicting $C_{l,t+1}$'s. This provides a "time series" interpretation of the model. For the statistician it ought to provide some explanation of temporal dependence; i.e., without such a term in the model, it

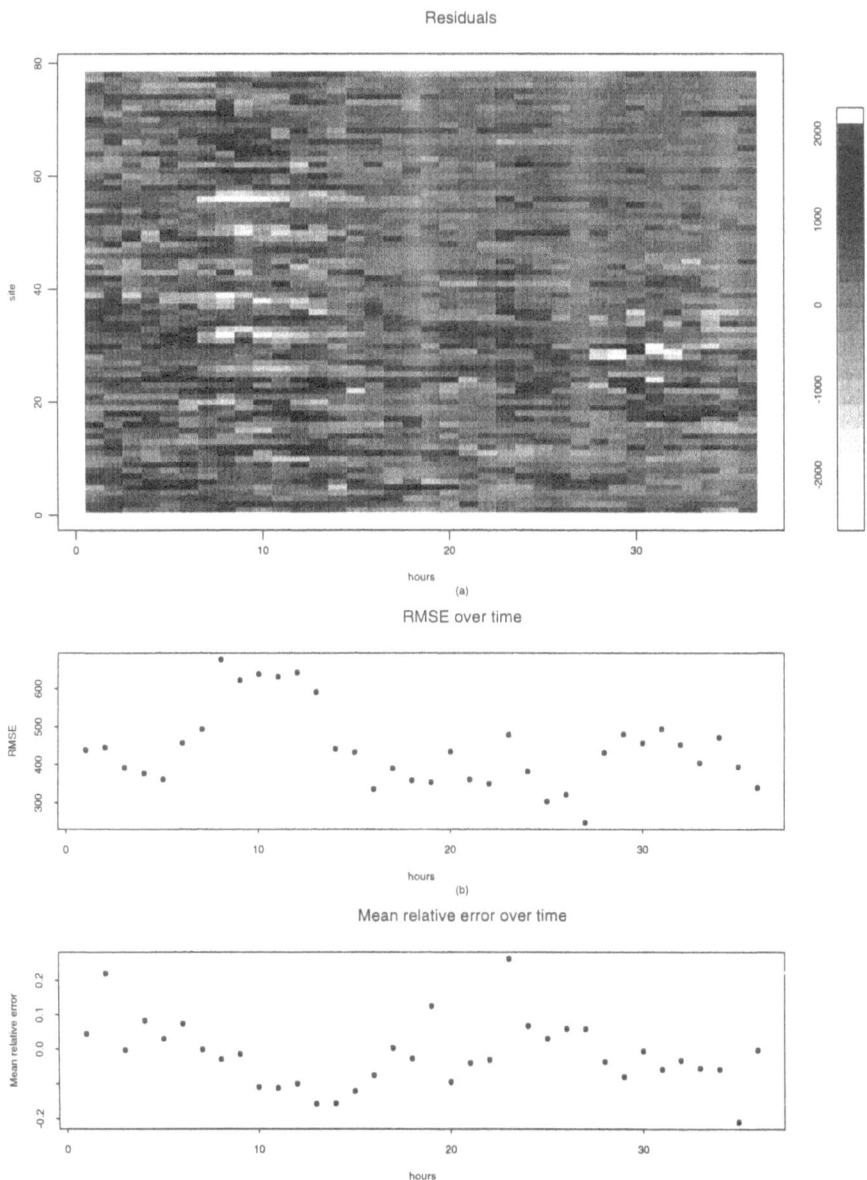

FIGURE 2. Cloud model fit using OLR: (a) image plot of the residuals; (b) plot of the RMSE of the sites over time; (c) plot of the mean relative error of sites over time.

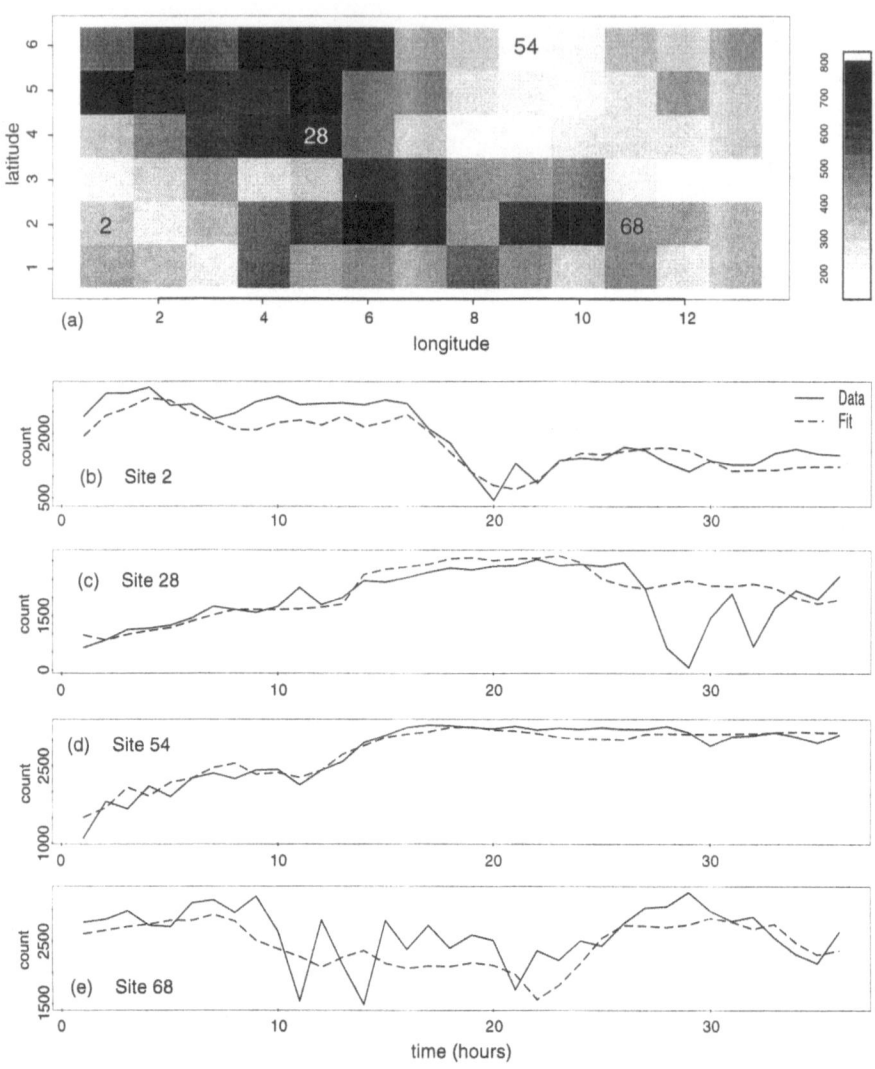

FIGURE 3. Cloud model fit using OLR: (a) image plot of RMSE over sites; (b) cloud cover (solid line) and predicted cloud cover (dotted line) for site 2 over 36 hours; (c) cloud cover (solid line) and predicted cloud cover (dotted line) for site 28 over 36 hours; (d) cloud cover (solid line) and predicted cloud cover (dotted line) for site 54 over 36 hours; (e) cloud cover (solid line) and predicted cloud cover (dotted line) for site 68 over 36 hours.

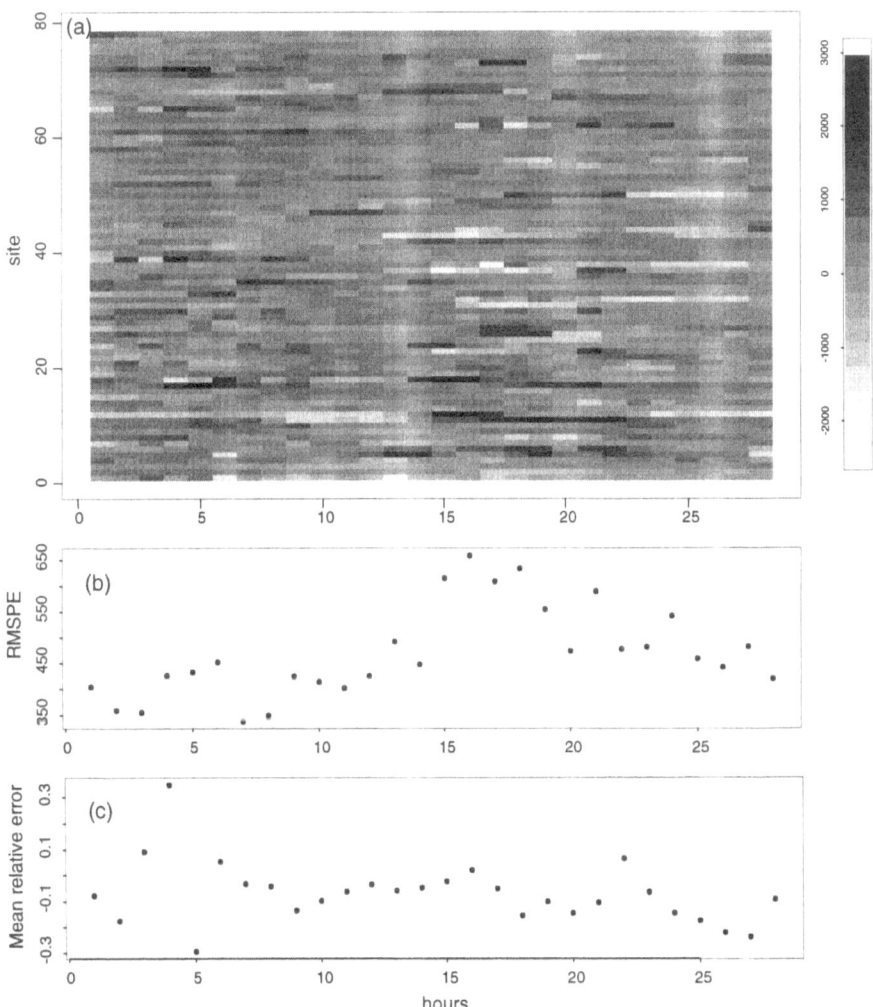

FIGURE 4. Cloud model prediction using OLR: (a) image plot of the residuals for the out-of-sample predictions; (b) plot of the RMSPE of the sites over time; (c) plot of the mean relative error of sites over time.

would be unlikely that model errors could even be approximately independent in time. For the cloud modeler we are building some degree of persistence into the model.

To a significant degree, cloud formation and evolution are controlled by

1. the background potential energy available in the atmosphere to enable convection (conversion to kinetic energy); and
2. the water vapor available for cloud formation and growth.

Hence, we consider three variables reflective of these controllers that would be available (computable) as output from a GCM. The variables are:

- the time rate of change of generalized convective available potential energy (DCAPE);
- Richardson's number (RICHN); and
- the column mean relative humidity (RHM).

We offer very brief descriptions of these variables. A more detailed discussion can be found in [Ema94]. Generalized convective available potential energy (CAPE) is a measure of the local vertical component of the moist available energy. Moist available energy is converted into kinetic energy by convection on the order of 1 hour. Richardson's number measures low level shear. It is the ratio of the velocity shear to the velocity associated with downdrafts. Relative humidity is the ratio between actual and saturation vapor pressure. (Saturation refers to a critical level of water vapor at which a phase change to liquid would occur. The saturation point depends on temperature and pressure.)

2.2.1 Data

Data regarding these predictor variables are available as model output from Florida State University (FSU). This model assimilated observational data from the TOGA COARE to produce output on a regular grid. The data are available on a 6-hourly time scale, i.e., four times/day at 0Z, 6Z, 12Z, and 18Z, on a 0.7° grid. In particular, the FSU data can be used to calculate our three variable predictors, DCAPE, RICHN, and RHM.

2.2.2 Model Fitting

First, note that in this analysis, we used the FSU grid (0.7°) in defining site (as opposed to the coarser GCM scale). Cloud cover is defined based on the original IR data analogously. Recall, the cloud cover data is on a 0.045° (5 km square) grid. Aggregating to the FSU grid implies that there are approximately 15^2 cloud count pixels in each site.

Let $C_{l,t}$ be the cloud cover at FSU grid site l at time t. Let $DCAPE_{l,t}$, $RICHN_{l,t}$, and $RHM_{l,t}$ be the values of the three explanatory variables, also at grid site l at time t. For each site (with full nearest-neighborhoods available, as before) l, the first-order time-lagged spatial nearest-neighbor model for $C_{l,t+1}$ is

$$C_{l,t+1} = f(C_{l,t}, \mathbf{DCAPE}_{\mathcal{N}(l),t}, \mathbf{RICHN}_{\mathcal{N}(l),t}, \mathbf{RHM}_{\mathcal{N}(l),t}, \theta) + e_{l,t+1}. \quad (7.13)$$

As before, the errors are all assumed to be mutually independent, zero mean Gaussians with common, unknown variance σ^2. Note that in the above model, $\mathbf{X}_{\mathcal{N}(l),t}$ in (7.9) is now a sixteen-dimensional vector: one for $C_{l,t}$, plus five (site l and its four nearest neighbors) times three explanatory variables. Recall that the sampling time for the FSU data is 6-hourly, but the time step of the prediction (and cloud data) is one hour. That is, $\mathrm{DCAPE}_{l,t}$, $\mathrm{RICHN}_{l,t}$, and $\mathrm{RHM}_{l,t}$ are constant in 6-hour batches. (Alternatively, one might consider interpolation schemes to produce the explanatory variables at an hourly rate; for brevity, we do not do so here.) Therefore, it is important to note for the model results in Section 2 that time is in 6-hourly increments. A time period of 16 6-hour increments spans four days.

The model in (7.13) was fitted separately to two different time periods, one convective (heavy clouding) and one nonconvective (relatively clear sky). (We explain our motivation for fitting these separate models at the end of the chapter.) The convective period was December 21–24, 1992; the nonconvective period was January 8–11, 1993. These classifications were chosen based on average values of OLR over the region [Che96a].

Figures 5–8 are a series of summary plots for both model fits and out-of-sample predictions for the two time periods. Each figure consists of five plots. In each case the first panel is an image plot of RMSE (or RMSPE) for all sites averaged over time. (White areas correspond to land, where no data are available. No analyses of any kind were performed here.) The second panel is a plot of RMSE (or RMSPE), averaged over sites, for each hour of the relevant period. The third panel is a plot of the corresponding mean relative error (see (7.12)). The last two panels display actual data and fitted (or predicted) values for two individual sites. The locations of the sites have been circled in the top panels and have been chosen for a contrast in RMSE (or RMSPE). Sites 200, 250, and 287 are in the upper left corner and Site 1125 is in the lower right corner of the grid.

Figure 5 is a summary of the fit to (7.13) with eight hidden units for the convective time period. We note an average of 22% RMSE. The area (lower right) of relatively small RMSE corresponds to the subregion that had the largest amount of cloud cover. Figure 5(b) displays the spatially averaged RMSE; the corresponding mean relative error in Figure 5(c) ranges between -0.34 and 0.35. The time series plots of observations and fitted values in Figures 5(d) and (e) indicate that the model is capturing how cloud cover evolves over time.

To examine the out-of-sample predictive performance, the fitted model was used to predict cloud cover count for another convective time period: January

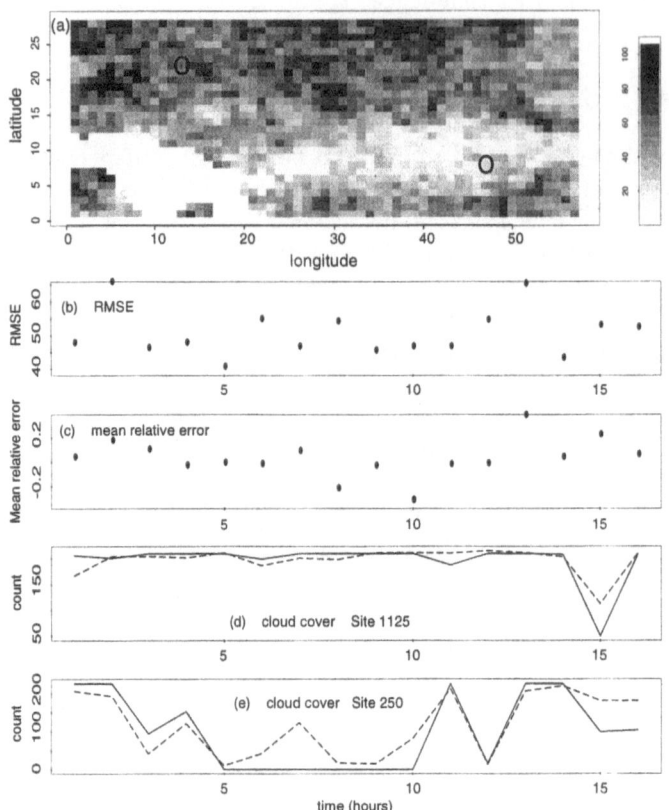

FIGURE 5. Cloud model fit: (a) image plot of the RMSE over sites for the convective period; (b) plot of the RMSE of the sites over time; (c) plot of the mean relative error of sites over time; (d) and (e) cloud cover (solid line) and predicted cloud cover (dotted line) for two selected sites, circled on (a).

26–28, 1994. Figure 6 is the corresponding summary plot for this period. The magnitude of RMSPE averages 24% over all sites and times. The predictive performance in time for two selected sites is shown in Figures 6(d) and (e). Site 200 is chosen instead of site 250, so that there is a contrast in RMSPE.

Figure 7 is a summary of the fit to (7.13) with eight hidden units for the nonconvective time period. The overall average RMSE is 17%, and the mean relative error ranges between −0.66 and 0.84. As before, observed and fitted values for two sites are plotted over time in Figures 7(d) and (e). Again, the plots indicate that the model is capturing how cloud cover evolves reasonably well. To examine out-of-sample prediction, the fitted model was used in another nonconvective time period February 4–6, 1994. Figure 8 is a corresponding summary plot of the fit to this period. The magnitude of the RMSPE has

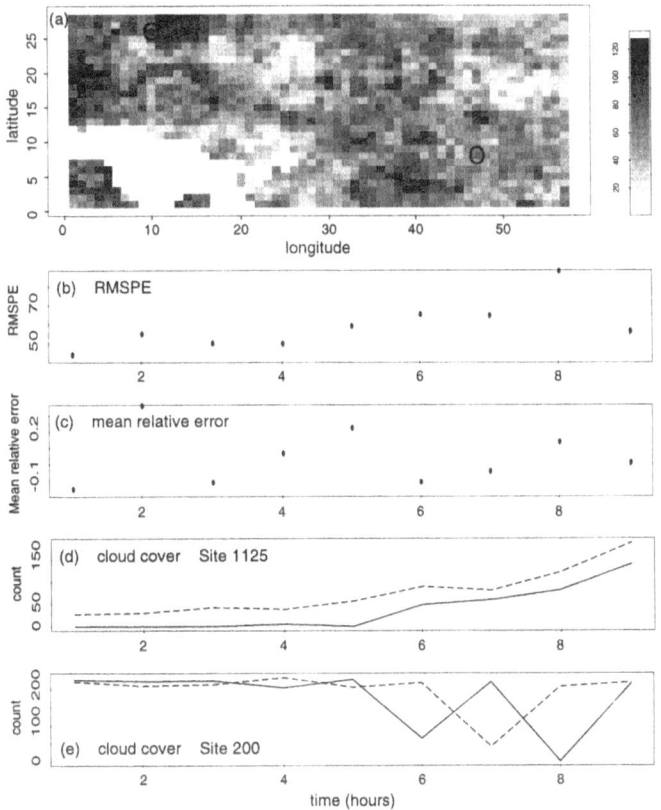

FIGURE 6. Cloud model prediction: (a) image plot of the RMSPE over sites for the out-of-sample convective period; (b) plot of the RMSPE of the sites over time; (c) plot of the mean relative error of sites over time; (d) and (e) cloud cover (solid line) and predicted cloud cover (dotted line) for two selected sites, circled on (a).

increased to approximately an average of 19% RMSPE over sites and 20% over time. Again, overall and site-specific performance is judged to be decent.

3   Simulation of Cloud Cover

Examination of the residuals for the time-lagged spatial nearest-neighbor model in (7.13) indicates that they are not correlated significantly over time, but there is significant spatial correlation. Adjusting for spatial correlation in fitting neural networks is extremely challenging and beyond the scope of this chapter. Since the fitted models appear to have reasonable out-of-sample predictive power, we offer the following enhancement that seeks to partially account for

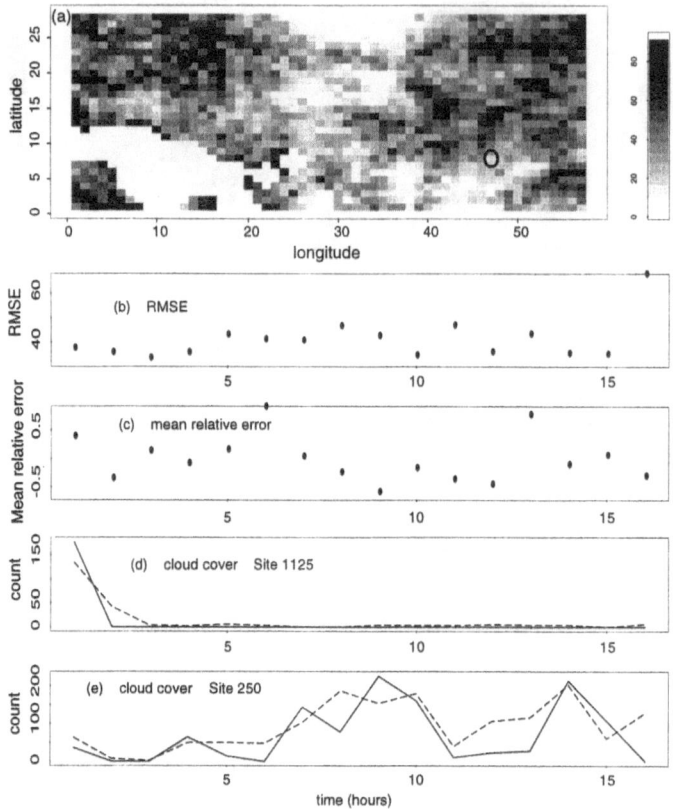

FIGURE 7. Cloud model fit: (a) image plot of the RMSE over sites for the nonconvective period; (b) plot of the RMSE of the sites over time; (c) plot of the mean relative error of sites over time; (d) and (e) cloud cover (solid line) and predicted cloud cover (dotted line) for two selected sites, circled on (a).

additional spatial structure.

The motivation for this analysis is to offer another check on the statistical model's ability to predict clouds. The check is to see how well the model simulates realizations of cloud cover patterns. The idea is to use the fitted model, but to predict out-of-sample site-wise cloud coverage, $C_{l,t+1}$, using

$$C_{l,t+1} = \hat{f}(C_{l,t}, \mathbf{DCAPE}_{\mathcal{N}(l),t}, \mathbf{RICHN}_{\mathcal{N}(l),t}, \mathbf{RHM}_{\mathcal{N}(l),t}) + Z(l,t+1), \quad (7.14)$$

where $\hat{f}$ is a fitted model as described in the previous section and the $Z(l, t+1)$ will be randomly generated errors possessing a prescribed spatial covariance structure. These new Z variates will be modeled using stationary Gaussian spatial processes.

First, note that by using the row–column index (i, j) notation for sites,

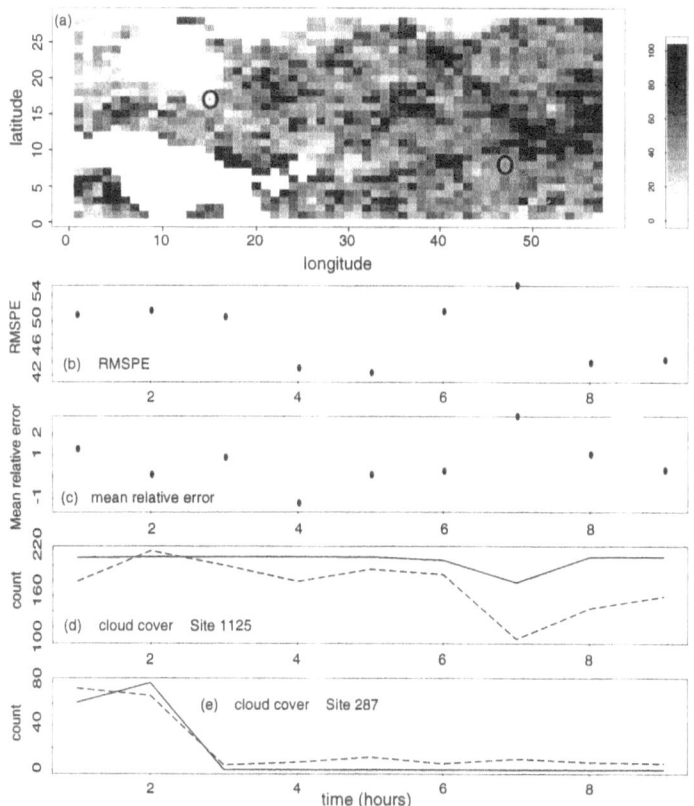

FIGURE 8. Cloud model prediction: (a) image plot of the RMSPE over sites for the out-of-sample nonconvective period; (b) plot of the RMSPE of the sites over time; (c) plot of the mean relative error of sites over time; (d) and (e) cloud cover (solid line) and predicted cloud cover (dotted line) for two selected sites, circled on (a).

we can readily define a Euclidean distance, $d(l, l') = \|l - l'\|$, between any two sites l and l'. For each time t, let $\{Z(l, t) : l = 1, \ldots, L^*\}$ be a random error process. We set the means of all $Z(l, t)$ to zero, and we will generate the $\{Z(l, t) : l = 1, \ldots, L^*\}$ processes mutually independently across time. Assume that the spatial covariances for the $\{Z(l, t) : l \in L^*\}$ do not vary in time (permitting suppression of dependence on time in the following definition), and are given by

$$\mathrm{COV}(Z(l), Z(l')) = \begin{cases} \sigma^2 \rho e^{-\|l - l'\|^2/\theta} & \text{if } l \neq l', \\ \sigma^2 & \text{if } l = l'. \end{cases}$$

Due to the exponential form, this is known as a Gaussian covariance function. The parameters of the covariance function, $\rho e^{-\|l - l'\|^2/\theta}$, were estimated by nonlinear least squares applied to the *original fitted residuals*. The estimated parameter values are $\hat{\rho} = 0.463$, $\hat{\theta} = 7.18$, and $\widehat{\sigma^2} = 2496.3$.

Overall, the residuals from the fitted model are reasonably viewed as approximately normally distributed. However, their patterns of variation are very dependent on the fitted values. This dependence is primarily due to the fact that cloud cover counts on a gridbox are constrained from 0 to 15^2. If the fitted values are close to the mean value of the cloud cover amount ($15^2/2$), then the residuals tend to behave as if they are normally distributed with mean zero. However, if the fitted values are near zero, the residuals will have a negatively skewed distribution because the model will tend to overpredict the true cloud cover amount. Similarly, if there is near total cloud cover, the model will tend to underpredict and the residuals will be positively skewed. To account for this dependence on fitted values, a beta distribution was fitted to the residuals corresponding to a range of fitted values. The fitted values were divided into 50 equally spaced bins over the range from 0 to 15^2. The goal is to generate random errors with the correct conditional mean and variance depending on the fitted value, and the correct exponential covariance structure. This is accomplished by adding a randomly generated residual from the appropriate beta distribution to a randomly generated variable with the correct spatial covariance to obtain $Z(l)$.

Realizations of cloud cover during the out-of-sample convective period were simulated as follows: starting at the first hour of the time period, a cloud cover realization was generated using (7.14) with a simulated vector 'of Z-variates. The result was then used in the model to simulate cloud cover (using a new, independently generated Z-vector) for the next time step, etc. Recall that convective data is only available 6-hourly, so that the parameterization is fit only to the convective data and cloud cover amount to predict the next hour time step. Note that to simulate these realizations at an hourly rate, interpolations of the convective explanatory variables were used. Figure 9 (a) and (b) is an image plot of the means of actual and simulated cloud covers for the out-of-sample convective time period. The spatial distribution indicates that the model is producing the dark areas, i.e., areas with relatively large amounts of cloud cover, as well as the light areas, fairly well. Figures 9(c)–(f) are time slices of cloud cover and simulated cloud cover at the second and sixth hour of the simulation. Note that the model has been iterated seven times to predict hour 2 and iterated 31 times to predict hour 6. Figures 9(g) and (h) are four cloud cover amount realizations for two selected sites.

4   Conclusions and Future Work

As noted in Section 3, despite our use of local spatial information in the development of the models based on (7.13), significant spatial correlation is present in the residuals. Two natural approaches suggest themselves. One is to use a

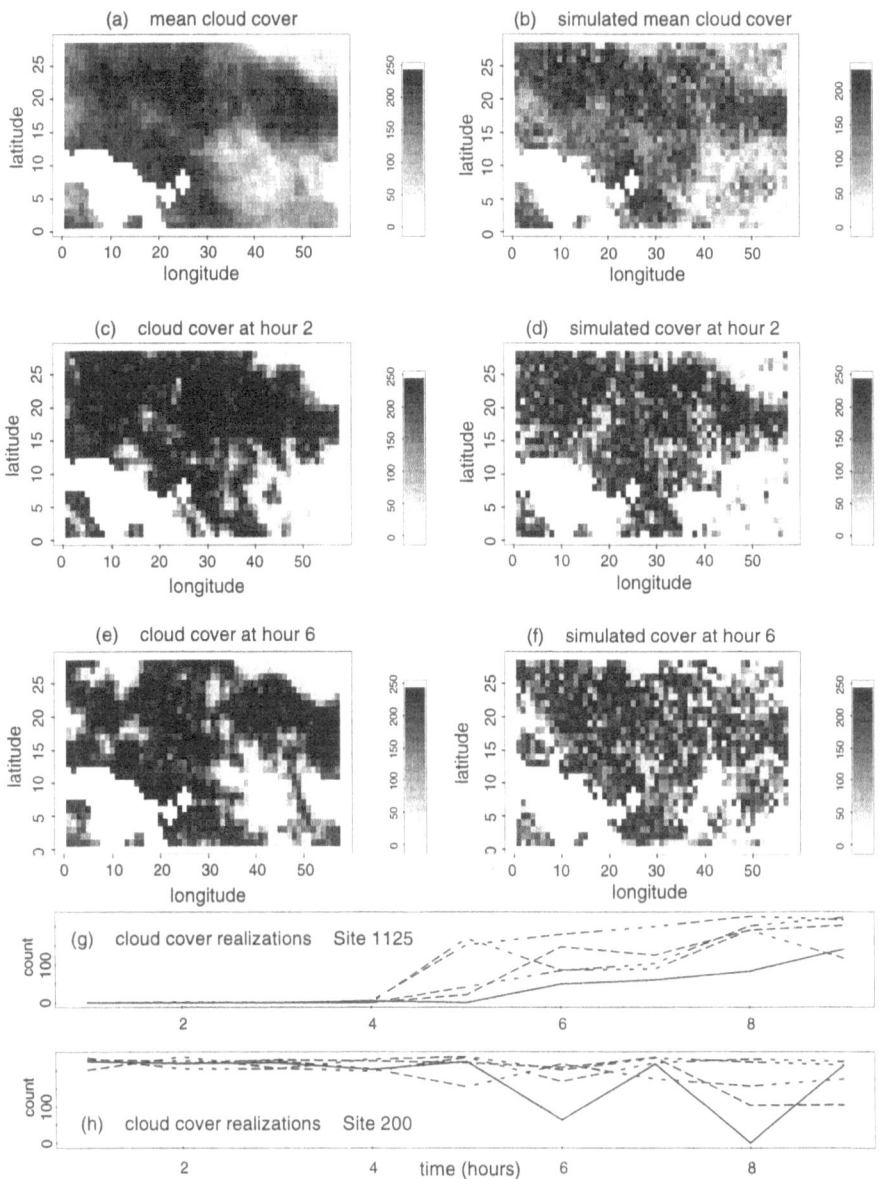

FIGURE 9. Simulated cloud cover: (a) mean cloud count over the out-of-sample convective period; (b) simulated mean cloud count for over the out-of-sample convective period; (c) cloud cover amount at hour 2; (d) simulated cloud cover amount at hour 2; (e) cloud cover amount at hour 6; (f) simulated cloud cover amount at hour 6; (g) and (h) four cloud cover amount realizations at selected sites.

spatially correlated errors formulation akin to that of Section 3 in the fitting of the neural network. (Statisticians will recognize switching from an *ordinary* least squares analysis to a *generalized* one.) However, this is an extremely difficult task when fitting complicated neural nets to large datasets. The second natural suggestion is to incorporate more complicated neighborhood structures and additional explanatory variables in model specifications. For example, to account for a climatological trend for advection from west to east in the tropical Pacific, we could include past cloud cover values of sites to the left of a modeled site. However, we must be prepared to pay a computational price in fitting more complicated models. Finally, clouds can be classified according to their temperature and a model that predicts different cloud types may prove to be even more useful.

While the development of *regional* models is of interest, our ultimate goal is the development of parameterizations for use in atmospheric GCMs. This would involve another order of magnitude in model fitting. However, other issues arise. There is strong reason to believe that the physics of cloud development and evolution depend strongly on the region studied. Hence, while neural networks may be capable of capturing cloud behavior sufficiently well for use in GCMs, it is likely that model parameters (regression coefficients), and indeed the form of model, ought to vary spatially. This actually may be a positive aspect in that it suggests development of regional models separately. On the other hand, it also suggests problems of linking such models at their boundaries as well as potential sensitivities to the exact definitions of regions.

Related to the notion of spatially varying models, there is a companion notion of state-dependent, time-varying models. In particular, in Section 2.2 we fitted separate models in convective and nonconvective time periods. As long as information relevant to determining such *regimes* is available in resolved GCM variables, such state-dependent models can be entertained. Indeed, current cloud parameterizations schemes do incorporate such regime indicators. Finally, notions of building complicated time-and-space varying models, including regime switching, can be readily formulated (though computation may still be difficult) from the Bayesian statistical viewpoint (see Lu and Berliner [Lu,99] and Wikle [Wik99a] (this volume) for examples and references). Such approaches would also enhance our ability to provide uncertainty measures associated with suggested parameterizations.

In summary, we believe we have made some progress in an extremely challenging and important problem in climate science. While we are encouraged, we also note that substantial interesting and exciting research awaits us.

Exploratory Statistical Analysis of Tropical Oceanic Convection Using Discrete Wavelet Transforms

Philippe Naveau
National Center for Atmospheric Research, Boulder,
CO 80307,USA

Mitchell Moncrieff
National Center for Atmospheric Research, Boulder,
CO 80307,USA

Jun-Ichi Yano
Cooperative Research Centre for the Southern Hemisphere
Meteorology Monash University, Clayton, Victoria 3168,
Australia

Xiaoqing Wu
National Center for Atmospheric Research, Boulder,
CO 80307,USA

1 Introduction

Convective cloud systems organized on scales ranging from a few kilometers to many hundreds of kilometers (mesoscales) are important for understanding the dynamics of weather and global climate. To model the atmospheric characteristics of these mesoscale cloud systems in the western Pacific, a variety of numerical models has been used in the past. Because of the very large size of the datasets, the statistical analysis of such simulated outputs can be very complex, and classical statistical tools are not always practical. The problem tackled in this chapter is the classification of a cloud system into three distinct regimes: highly organized squall-lines, less organized nonsquall cloud clusters, and scattered convection (a detailed description of these regimes can be found in Section 2.1). In interpreting the numerical output according to these three regimes, we will draw on recent statistical methodology for image analysis.

A classification procedure should be efficient. It must take into account the double-periodicity of the modeling region and must provide a physical interpretation. We will present an algorithm based on discrete wavelet transforms that satisfies these required conditions (basic wavelets properties are reviewed in Section 3). One reason for the success of the wavelet analysis is that it retains the spatial structure of the cloud at different scales. By selecting wavelet coefficients, we were able to keep the most important characteristics of the cloud, i.e., the ones that define the different regimes.

1.1 Atmospheric Motivation

In this section, we present some of the atmospheric considerations that motivated this research.

Clouds influence the heating and cooling (energetics) of the atmosphere in two primary ways:

- Clouds play an important role in the atmospheric water cycle, i.e., release of latent heat by condensation and removal of liquid water by precipitation.
- By scattering, absorption, and emission of solar and terrestrial radiation, clouds strongly affect the radiation budget (the *Earth-atmosphere* system has to maintain a balanced heat budget).

Because mesoscale clouds are large coherent structures, their study is an essential component for our understanding of the atmospheric water cycle and the radiation budget. Also, organized cloud systems have a direct dynamical effect through vertical motion of the atmosphere (momentum transport), a process that is only beginning to be quantified. Mesoscale clouds can generate very large and strong squall-lines that are responsible for severe weather. Hence, a quantitative description of squall-lines will also lead to a better understanding of weather impacts. The reader can find a more detailed description of mesoscale systems in the review article by Moncrieff [Mon95].

1.2 Statistical Motivation

The increase in computer power in the last decades has led to better and more complete simulated cloud fields. As a complement to this end, atmospheric scientists need powerful statistical tools to explain, summarize, and improve the outputs of numerical models. A statistical analysis of simulated clouds is, by itself, a challenging task. Our motivation was to provide new and simple statistical procedures that can be applied to large datasets and help understand mesoscale clouds. These procedures are based on the statistical characteristics of wavelets. The recent development of wavelet bases based on multiresolution analysis has been applied to a wide variety of theoretical and practical problems. Thus, we believed these models would also be useful in this context. In the field of statistics, Dohono and Johnstone [Don95] applied wavelet transforms to nonparametric regression by thresholding wavelet coefficients. In the area of geophysics, Yano et al. [Yan99] used wavelet transform to decompose a numerical realization of tropical oceanic convective cloud systems into two primary components, based on the vertical shear of the horizontal wind (the rate of change of the horizontal wind vector in the vertical direction). In our case, we develop a statistical technique based on wavelets to classify different regimes.

1.3   Objectives

Our main intentions are twofold: first, to provide a basic introduction to the statistical problems encountered when studying mesoscale clouds, and second, to show the potential of wavelet transforms for geophysical processes and their statistical analysis. To illustrate these two points, we focus on the problem of identifying (classifying) the main components of a mesoscale cloud system, i.e., the evolution of a population of convective clouds into different organized regimes. By convection, we mean the vertical currents made visible by clouds.

The chosen procedure will have to:

- be fast to compute (because of the high dimensionality of our dataset);
- account for the double-periodicity of our images;
- detect the presence or absence of squall-line regions;
- classify cloud systems into one of three distinct physical regimes; and
- give results that are physically interpretable.

2   Description of the Dataset

The entity of interest is the tropical oceanic mesoscale convective system. Squall-lines are the paradigm of organized convective systems. For a detailed description of mesoscale convective systems and their numerical simulation, see [Cot89]. The dataset is the output from a large numerical simulation and is constituted of 504 temporal snapshots taken every 20 minutes on a three-dimensional spatial domain of size $400 \times 400 \times 16$ (km) and on the specific grid described by Table 8.1. A complete description of this dataset can be found in Wu and Moncrieff [Wu,96].

TABLE 8.1. Grid characteristics.

Direction	Number of grid points	Resolution
Vertical	33	0.5 km
x-axis	128	3.125 km
y-axis	128	3.125 km

Because we want to moderate the computational cost, the wavelet transform appeared as a natural candidate (a very fast algorithm developed by Mallat [Mal89] exists for computing wavelet coefficients for a dyadic number of observations, i.e., 2^K for some integer K). Originally, the data were not available on a grid of dyadic points, so they were interpolated in order to be available on the grid described in Table 8.1.

Another interesting aspect of our dataset is that it is not composed of "real" data (i.e., physical phenomena measured by instruments), but it is simulated from a numerical model. For a statistician, this implies an additional error. The main reason we study numerically simulated (synthetic) data is that real atmospheric data of adequate resolution are not available. Fortunately, atmospheric numerical models now produce realistic simulation of cloud systems because they are based on physical laws, written as a set of differential equations, that describe the physics of the motion and the thermodynamics of air parcels. Although it is not possible to analytically solve these basic equations, numerical simulations emulate the macroscopic behavior of the atmosphere, and the validity of numerical models has been extensively tested by the atmospheric community. One constraint of these numerical models is that some artificial boundary conditions must be imposed. This disadvantage of numerical models turns out to be fortuitous when using wavelets, since boundary conditions also have to be fixed for wavelet analysis. In our case, the numerical model is doubly periodic in space at fixed altitude, i.e., each horizontal plan of a snapshot at a specific altitude z is periodic in the x-direction and the y-direction. So the wavelets naturally inherit the same boundary condition.

Due to the enormous size of the complete dataset, only a few snapshots have been processed at the present time. These snapshots are representative of the three different cloud regimes under study. As future work, we plan to analyze a much larger number of snapshots to validate our statistical model.

TABLE 8.2. Variables of interest.

u	East wind velocity
v	North wind velocity
w	Vertical wind velocity
θ	Potential temperature
q_v	Moisture mixing ratio (g/kg)
q_c	Cloud water (g/kg)
q_r	Rain water (g/kg)
q_i	Ice water (g/kg)
t	Total water condensate ($t = q_c + q_r + q_i$)
k	Kinetic energy

The variables listed in Table 8.2 characterize the cloud system, and are available at each point on our grid. The moisture mixing ratio q_v is equal to

$$q_v = \frac{grams\ of\ H_2O\ vapor}{grams\ of\ dry\ air},$$

the quantity k is proportional to the kinetic energy of a parcel of air and is defined by

$$k = \frac{u^2 + v^2}{2},$$

where u (respectively, v) is the east (respectively, north) wind velocity.

2.1 Cloud System Regimes

As mentioned earlier, each snapshot of the data belongs to one of three cloud system regimes: *squall-line, nonsquall clusters*, and *scattered convection*. This decomposition into three regimes is important, because they reflect important physical behaviors. The squall-line regime, also called an organized mesoscale convective system, occurs in an environment characterized by a strong vertical wind shear. The nonsquall cluster regime occurs in moderate wind shear and features a less coherent spatial structure than the squall-line. Scattered convection typically occurs under weak shear conditions. Figure 1 shows three different total water condensate fields (see Table 8.2) for these different regimes, and a horizontal view at an altitude of 5 km is presented in Figure 2.

While wind shear is an important dynamic control on the organization of convection, the generation of convection is the result of the motion of large air masses and the fluxes of heat from the ocean. This is termed *large-scale forcing*. Figure 3 shows the large-scale forcing during the time of the experiment, i.e., the conditions applied to the numerical model during the simulation. The continuously-changing combination of wind direction, air temperature, and moisture content *forces* the model to create the different kinds of regimes mentioned above. In the present study, the large-scale forcing is obtained from observational measurements made during 1–7 September 1974 in the Global Atmospheric Research Program (GARP), Atlantic Tropical Experiment (GATE) conducted over the eastern tropical Atlantic off the coast of west Africa.

As one might expect, the definition of each regime is not mathematically well defined, and there are time periods where no "typical" regimes are observed, or a combination of two regimes occurs.

As seen in Figure 3, the nonsquall cluster develops under the strong large-scale forcing and weak low-level shear (day 2) and is characterized by a circular type of structure, called cumulus congestus. Various types of clouds such as deep and shallow clouds and small cloud bands can be found in the plot of the condensate field (see Figure 1). Squall-lines occur under the moderate forcing and strong low-level wind shear (day 4). The interesting features are the pronounced vertical motion of air in the leading edge of the system, and extensive anvil clouds (about 200 km wide) to the rear. Scattered convection is a response to the weak forcing condition (day 7) and has a southwest–northeast oriented line structure. The scattered convection is mainly controlled by the

(a) *Nonsquall Cluster*

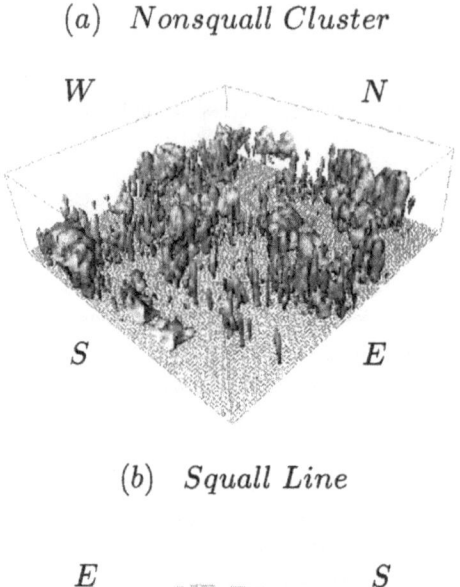

(b) *Squall Line*

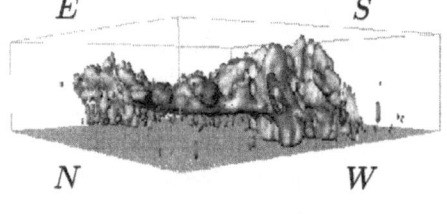

(c) *Scattered Convection*

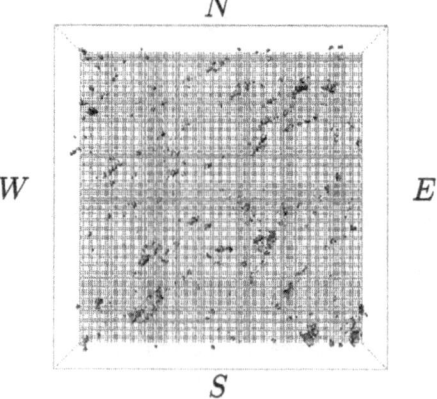

FIGURE 1. Snapshots of the total condensate field with the isosurface 0.1 g/kg.

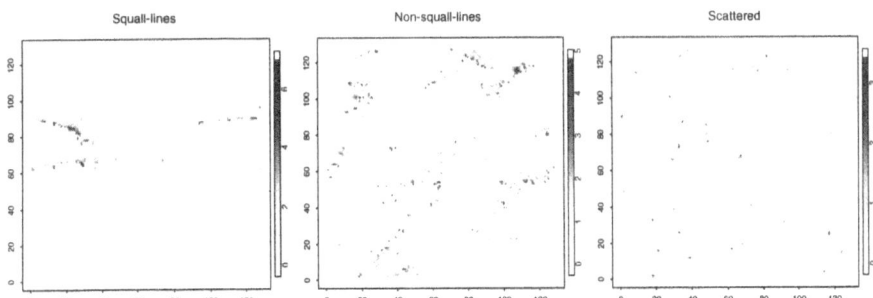

FIGURE 2. Total water condensate fields for three different cloud regimes at altitude 5 km.

low-level wind field. The three distinctive convective regimes are described in more detail in [Wu,96] and [Yan99]. [1]

3   Discrete Wavelets

Over the last decade, wavelets have been widely applied in different disciplines. One reason for the success of wavelets is that they offer a powerful and efficient tool for data-reduction. Wavelets are basis functions that are organized at multiple levels of resolution. In the next two subsections, we present a quick review of some properties of wavelets, in one and two dimensions.

3.1   Wavelets in One Dimension

Classical regression techniques in statistics might represent a function as a specific degree polynomial or as the sum of sine and cosine functions at different frequencies. A wavelet analysis corresponds to a particular type of series expansions. The wavelet basis is composed of orthonormal functions $\psi_{j,k}$:

$$\psi_{j,k}(t) = 2^{j/2}\psi(2^j t - k) \qquad \text{for } j, k \text{ integers,}$$

where the function ψ is a fixed function in $L^2(R)$, called the *mother wavelet*. Note the double indexing includes both scaling and translation in this family. A rigorous framework concerning the construction for such orthonormal basis can be found in [Mey93] and [Dau92]. Hence, if $f(t)$ is a function in $L^2(R)$, we have the following expansion:

$$f(t) = \sum_{j=-\infty}^{\infty} \sum_{k=-\infty}^{\infty} \langle f, \psi_{j,k} \rangle \psi_{j,k}(t),$$

---

[1]Visualizations of the simulation can be found at   http://www.scd.ucar.edu/vg/GATE/GATE.html. The simulation depicts the formation and evolution of cloud systems over a seven-day period in the eastern tropical Atlantic region.

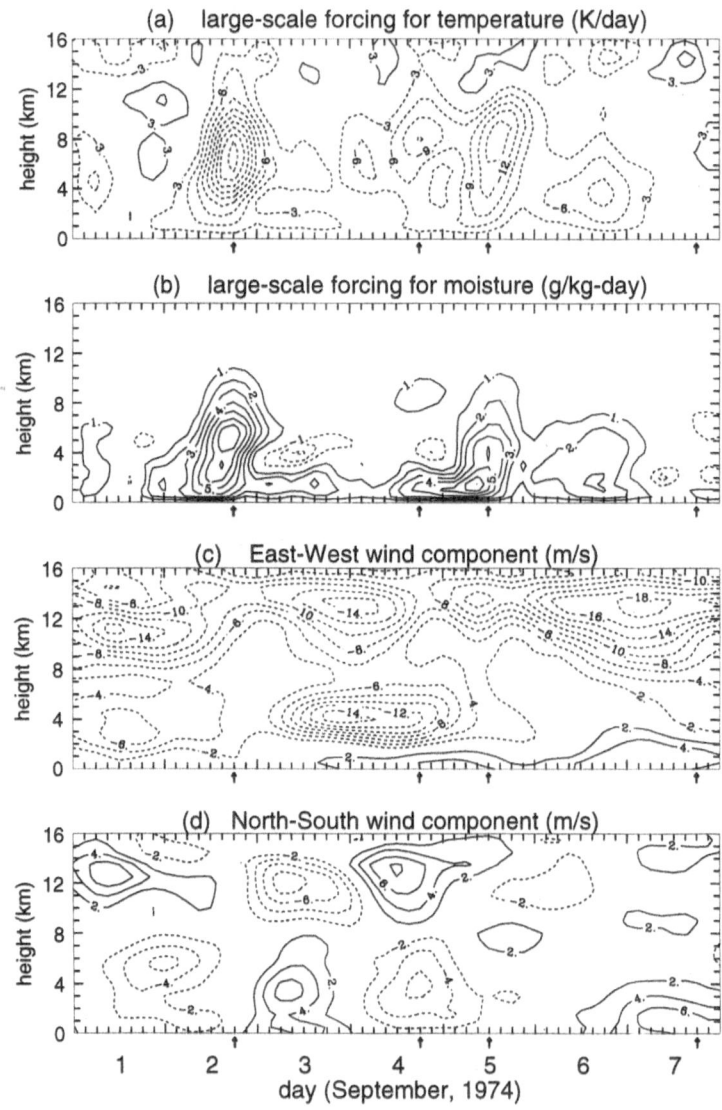

FIGURE 3. Evolution of the large-scale forcing for the temperature (panel (a)), the water vapor mixing ratio (panel (b)), the east–west wind component (panel (c)), and the north–south wind component. Contour interval are 3 K day^{-1}, 1 g/kg day^{-1}, and 2 m s^{-1}.

where $\langle .,. \rangle$ is the classical inner product defined by $\langle f,g \rangle = \int f(x)g(x)\,dx$.

In contrast to Fourier transforms, the wavelet transform uses different basis functions to represent the signal at different locations. A rigorous relationship between the local behavior of a signal and the local behavior of its wavelet coefficients exists. For instance, if the function is locally smooth around a certain point, then the corresponding wavelet coefficients will remain small, and if the function contains a discontinuity, then the corresponding wavelet coefficients will be much larger.

The *multiresolution analysis* is one fundamental property of wavelets. At each step of increasing (decreasing) resolution j, a finer (coarser) approximation of the original signal is created. Moving from coarser to finer approximation, or from finer to coarser approximation, is known as the "zoom-in zoom-out" feature of multiresolution analysis.

In this particular chapter, we will use the Daubechies wavelets of order 8, since they give good results for our specific situation. These orthonormal wavelets are compactly supported. For more information about these wavelets, see [Dau92].

3.2 Discrete Wavelet Transform in Two Dimensions

In this subsection, two-dimensional wavelets[2] are described. There are various ways to construct two-dimensional wavelets from one-dimensional wavelets. For our purposes, we use the extension based on a special tensor product that allows only a single resolution index j. For a formal presentation of this construction, see Chapter 10 of [Dau92]. The two-dimensional mother wavelets are defined by the following equations:

$$\begin{aligned}
\psi^h(x,y) &= \phi(x)\psi(y), \\
\psi^v(x,y) &= \phi(y)\psi(x), \\
\psi^d(x,y) &= \psi(x)\psi(y),
\end{aligned}$$

where the indices h,d,v indicate a specific direction (horizontal, diagonal, or vertical). After dilating and translating for a fixed dilation index, say j, we obtain

$$\psi^m_{j,k_1,k_2}(x,y) = 2^{j/2}\psi^m(2^j x - k_1, 2^j y - k_2) \qquad \text{for} \quad m = h,d,v.$$

By inheritance of the properties of the scaling function ϕ and the mother wavelet ψ, it follows that

$$\{\psi^m_{j,k_1,k_2},\ m = h,d,v \text{ and } j,k_1,k_2 \text{ integers}\}$$

---

[2]We use the algorithm as implemented in S-PLUS by Nason [Nas93].

is an orthonormal basis for $L^2(R^2)$. Any two-dimensional signal $f(x, y)$ that is square integrable over the real plane can be written as

$$f(x, y) = \sum_{m=h,d,v} \sum_j \sum_{k_1,k_2} w^m_{j,k_1,k_2} \psi^m_{j,k_1,k_2}(x, y).$$

Mallat [Mal89] noted that the three sets of wavelet coefficients correspond to a specific spatial orientation. For example, Figure 4 displays a test image with its wavelet coefficients. The vertical details in Figure 4 are represented in the lower right-hand portion of the plot, the horizontal characteristic can be seen in the upper left-hand corner, and the diagonal elements are in the top right-hand corner. This spatial phenomenon is observed for each level of resolution. Figure 5 shows three reconstructions of the test image from Figure 4 by using only the "diagonal" (respectively, "horizontal" and "vertical") wavelet coefficients. From Figure 5, it is very clear that each of three reconstructions preserves a direction. This property will play an essential role in our statistical procedures. The idea is quite simple: A squall-line regime, which by definition has a special orientation, should be more prominent in one of the three directions. In the next section, we will develop this idea in more detail.

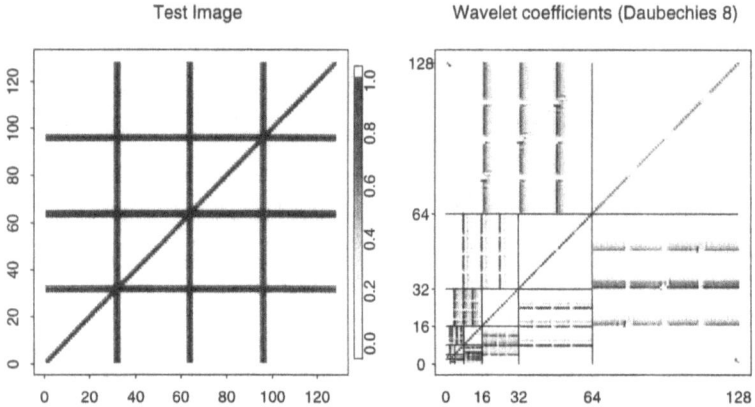

FIGURE 4. A test image and its wavelet coefficients.

Note that the terms, "diagonal," "horizontal," and "vertical," as used in the wavelet literature to indicate (mathematical) directions in a two-dimensional image, require clarification here. In geophysics, the term "vertical" is usually used to denote the vertical component of a three-dimensional, physical, co-ordinate system. In the following sections, the distinction between the two definitions of "vertical" should be clear according to the context.

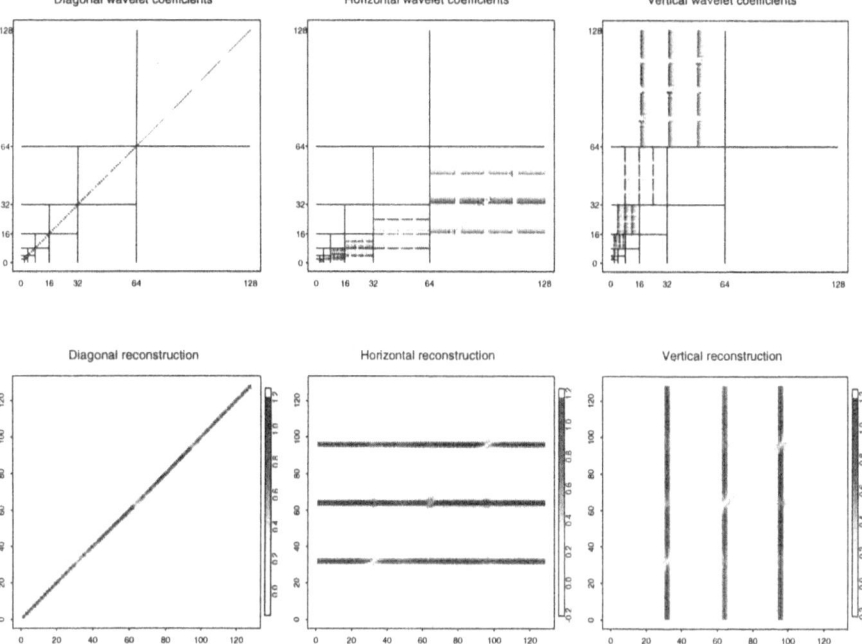

FIGURE 5. Wavelet coefficients and the respective reconstructions.

4   Statistical Study of Cloud Systems

4.1   Detecting Squall-Lines

In Section 3.2, we noted that each wavelet coefficient corresponds to a specific orientation, a particular spatial scale, and a particular location. For example, suppose one wants to measure the error between one of the reconstructed images (horizontal, diagonal, or vertical), say $\{T_{r,s}^{(m)}(z)\}$, with $m = h, d, v$ and the original image of total water condensate at a specific altitude z, say $\{T_{r,s}(z)\}$. Then, a natural statistic, called the *reconstruction error*, is

$$E_T^{(m)}(z) = \sum_{r=1}^{128}\sum_{s=1}^{128}(T_{r,s}(z) - T_{r,s}^{(m)}(z))^2 \qquad \text{with} \qquad m = h, d, v. \qquad (8.1)$$

To formally define the variable $\{T_{r,s}^{(m)}(z)\}$, we express its wavelet coefficients, say $\{w_{j,k_1,k_2}^{n,m}\}$ for $n, m = h, d, v$, in terms of the wavelet coefficients of the original image, say w_{j,k_1,k_2}^{m}. The relationship between these two sets of wavelet coefficients is

$$w_{j,k_1,k_2}^{n,m} = \begin{cases} w_{j,k_1,k_2}^{m} & \text{if } m = n, \\ 0 & \text{if } m \neq d. \end{cases}$$

Of course, we can compute these reconstruction errors for any variable in Table 8.2. If no particular direction is present in the original image, such as the scattered and the nonsquall cluster regimes, then the three reconstruction errors should be approximately equal. In contrast, the squall-line regime, which emphasizes a particular orientation, should generate a large discrepancy between one specific direction and the other two.

4.2 The Orientation Problem

Before presenting applications based on these reconstruction errors, one may ask what happens if the squall-lines are not perfectly horizontal, vertical, or diagonal. Hence, the validity of the reconstruction error criterion should be examined under different orientations. In Figure 6, we apply the reconstruction error criterion to test images like the one on the left side of Figure 6. Each image test contains a line with a different angle, ranging from 0° to 90°. Because of the symmetry, the value of the error criterion for an angle greater than 45° can be easily deduced from this plot. The background for each test image is white noise with a standard deviation equal to 0.2.

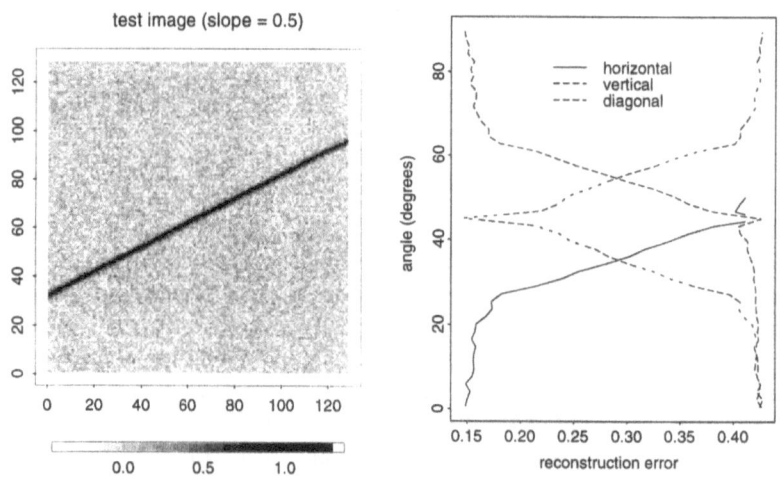

FIGURE 6. Reconstruction error for lines with different slopes.

From Figure 6, we see that the power of discrimination between the three principal directions is strong in the sense that the horizontal error criterion is able to identify any line with an angle between 0° and 30°. The same is true for an image with a vertical component, i.e., an angle between 60° and 90°. If the angle is between 30° and 60°, and not close to 45°, then the distinction between the three principal directions is not as clear-cut. One easy solution is to rotate the image by 45°. Then the angle can be clearly identified as either a horizontal or a vertical component.

4.3 Examples

To illustrate the efficiency of the reconstruction error criterion with real data, we plot, in Figure 7, the reconstruction error of the kinetic energy, $E_K^{(m)}(\cdot)$, at different altitudes for three different snapshots. These particular snapshots can be considered as archetypes of the three different regimes (see Figure 2). The altitude in this graph ranges from 1 km to 8 km since the interesting characteristics of the cloud are between these two altitudes. From Figure 7, we notice that no significant differences exist between the diagonal, horizontal, and vertical error reconstruction for the scattered convection and the non-squall cluster regime. In particular, the error for the scattered convection is small above 4 km. This is confirmed by the atmospheric knowledge that scattered convection is mostly limited to altitudes below 4 km. In comparison, the horizontal reconstruction for the squall-line regime is very different from the vertical and diagonal ones. This phenomenon starts around 2 km altitude, increases steadily to around the 4.5 km peak, and slowly decreases with higher altitudes. Hence, we deduce that horizontal squall-lines must exist in this image, and that they are particularly strong between 3 km and 6 km. This is in accordance with the original image.

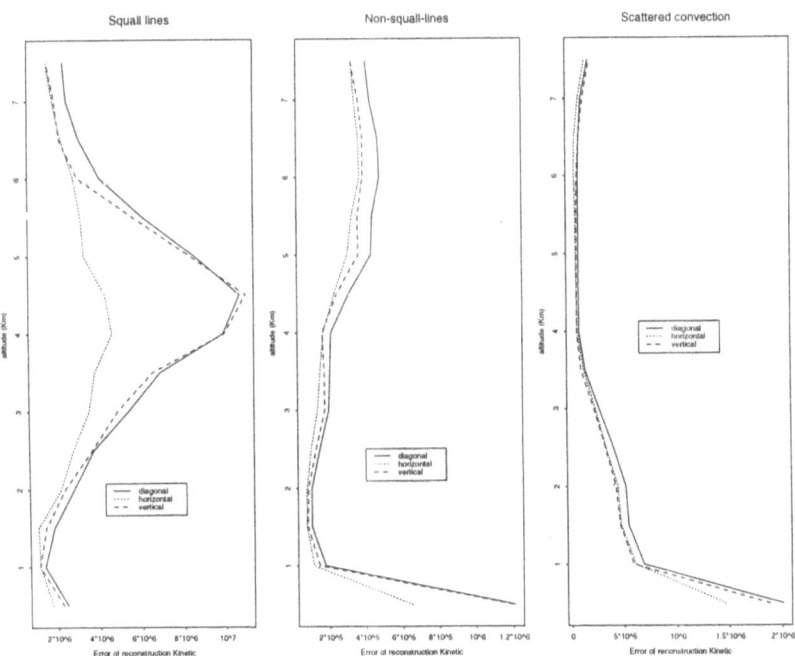

FIGURE 7. Reconstruction error for the kinetic energy, $(u^2 + v^2)/2$, as a function of the altitude.

Hence the reconstruction error appears to be a valid criterion to distinguish the squall-line regime from the scattered convection and also the nonsquall cluster regime. But this criterion is not efficient to discriminate between the scattered convection and the nonsquall cluster regime. In the next subsection, we will present another tool based on the isotropic character of the scattered convection, and the multiscale properties of the wavelet decomposition.

4.4   Identifying the Scattered Cloud Regime

The scattered cloud regime can be described as an isotropic state at any scale, and no specific orientation is favored either globally or locally. At large scales, this is also true for the nonsquall cluster regime, but when one looks at smaller scales, the clusters violate overall isotropy. Since wavelets are basis functions that can represent an image at different levels of detail, it is possible to construct statistics based on wavelets that take advantage of this multiresolution property. Concerning computational cost, we have already computed the wavelet coefficients to get the error reconstruction (see (8.1)). So, the proposed algorithm to detect the scattered cloud regime will not significantly increase the computational cost.

To take advantage of the multiresolution of the wavelets, we need to introduce a new error criterion that is defined at *each level of resolution, j_0*:

$$E_{T^{(j_0,m)}}(z) = \sum_{r=1}^{128}\sum_{s=1}^{128}(T_{r,s}(z) - T_{r,s}^{(j_0,m)}(z))^2 \text{ with } m = h,d,v \text{ and } j_0 = 1,...,6,$$

$$(8.2)$$

where $T_{r,s}(z)$ is the original field and $T_{r,s}^{(j_0,m)}(z)$ is a reconstructed image obtained by only changing the original wavelets coefficient at the resolution level j_0. More precisely, the wavelets coefficients of $T_{r,s}^{(j_0,h)}(z)$, $w_{j,k_1,k_2}^{h,m}$, are defined by

$$w_{j,k_1,k_2}^{j_0,h,m} = \begin{cases} w_{j,k_1,k_2}^{h} & \text{if } j \neq j_0, \\ w_{j,k_1,k_2}^{h} & \text{if } j = j_0 \text{ and } m = h, \\ (w_{j,k_1,k_2}^{d} + w_{j,k_1,k_2}^{v})/2 & \text{if } j = j_0 \text{ and } m \neq h. \end{cases}$$

In other words, we keep all original wavelet coefficients for any level of resolution different from j_0. For the level j_0, we replace the diagonal and vertical wavelet coefficients by an average. If the image is isotropic, the difference between the original image and the reconstructed ones should be approximately the same (relative to the number of coefficients changed, i.e., on a logarithmic scale). Indeed, the values of the wavelet coefficients should be approximately similar at each resolution level for an isotropic image.

Figure 8 illustrates this procedure for the three different total condensate fields at altitude $z = 5$ km of Figure 2. Each row corresponds to a different regime (respectively, nonsquall cluster, squall-line, and scattered), and each

column, respectively, represents the vertical, horizontal, and diagonal components. In each frame, the graph

$$(j_0, \log(E_{T^{(j_0,m)}}(z))), \qquad j_0 = 1, ..., 6, \qquad z = 5 \qquad \text{and} \qquad m = h, d, v,$$

is plotted. Hence, we observe that the logarithm of the error defined by (8.2) for
the scattered case (as expected for a quasi-isotropic image), is a linear function
of the resolution level. The estimated slopes are of the same order. This may
indicate that the scattered regime possesses a scale-invariant property.

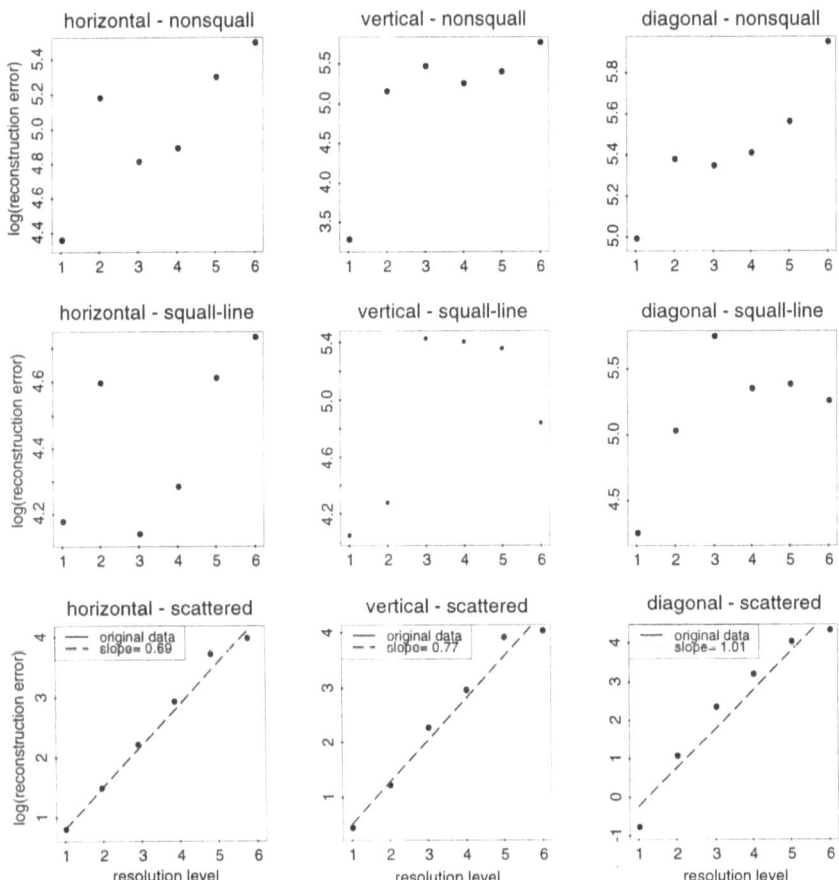

FIGURE 8. Identification of the scattered convection.

In contrast, linearity vanishes for the nonsquall cluster and squall-line regimes.
The departure from linearity is most prominent for the squall-line regimes,
because the structure in this regime is the opposite of the isotropy. Hence,
a simple regression test on the log error will identify the cases of scattered
convection.

5   Conclusions and Future Work

We have described a simple statistical algorithm for detecting the different regimes present inside a mesoscale convective cloud system. This algorithm was based on wavelet analysis. More specifically, using patterns in the basis functions with different orientations at different levels of resolution has enabled us to detect some characteristic signatures for the three different regimes. Data fields simulated by numerical models have shown the applicability and usefulness of the proposed statistics.

Future statistical analysis is needed to model the influence of large-scale forcing on different fields, such as the total water condensate field. A stochastic model to simulate the three different cloud regimes would be of prime interest to the atmospheric science community. Such a stochastic model could also be based on wavelets, and could incorporate some ideas presented in this chapter.

Predicting Clear-Air Turbulence

Claudia Tebaldi
*National Center for Atmospheric Research, Boulder,
CO 80307, USA*

Doug Nychka
*National Center for Atmospheric Research, Boulder,
CO 80307, USA*

Barbara Brown
*National Center for Atmospheric Research, Boulder,
CO 80307, USA*

Bob Sharman
*National Center for Atmospheric Research, Boulder,
CO 80307, USA*

1 Introduction

High-altitude, *clear-air turbulence* (CAT) presents an important forecast problem for aviation. It may cause injuries to passengers and crew, structural damage, an increase in fuel consumption, and in some circumstances fatal accidents.

CAT is a small-scale meteorological phenomenon whose triggering mechanisms are still under study. Though often linked to vertical wind shear, for example, at the boundaries of the jet stream, its complete causes are still unknown. Most attempts to forecast CAT have used indices based on synoptic conditions conducive to turbulence derived from the output of operational forecast models. Unfortunately, CAT is not resolvable at the grid spacing of such models. All the indices computed to determine the likelihood of CAT have proven to be only weakly correlated to actual CAT occurrences. To our knowledge, no effort has been devoted to rigorously evaluate the effectiveness of such indices and combine their strengths in a multidimensional and statistical model.

Our studies attempt to build an effective regression model from observational data, recorded on US Airways aircraft along their regular flight path, and using a suite of indices as predictors. Such indices are computed from the output of the Rapid Update Cycle (RUC-60) model, in operational use by the National Oceanic and Atmospheric Administration [Ben98]. In the following sections we give some background on the computation of CAT indices, statistical methods for flexible discriminant analysis and, finally, an evaluation of our approach for several study periods.

The reader is referred to the *web companion* for specific datasets and software related to the case studies in this chapter.

2   Indices Derived from the RUC-60 Model

The RUC-60 system produces three-dimensional analyses and short-range predictions every 3 hours by combining information from a large-scale meteorological observational network (rawinsonde, aircraft, profiler, and surface observations) with a background field (usually from the previous model output), producing weather forecasts over the continental United States for the next 3-hour period.

The RUC-60 horizontal domain covers the contiguous 48 United States and adjacent areas of Canada, Mexico, and oceans with a grid whose boxes are, on average, 60 km × 60 km in the horizontal dimension, and 1 km in the vertical dimension. The total vertical domain is resolved by 11 layers extending from roughly 3 km to 18 km above the surface, through an irregular vertical spacing based on the local thermal structure and its seasonal evolution.

The quantities analyzed and predicted by the RUC-60 system, at each forecast time, are:

- pressure;
- Montgomery stream function[1];
- virtual potential temperature[2];
- condensation pressure (equivalent to the lifting condensation level[3]);
- the horizontal wind components (u and v) relative to the grid.

From the 6 quantities listed above, 15 indices are computed as turbulence diagnostic indicators:

- Ellrod's number [Ell92b];
- Turbulent Kinetic Energy: TKE and TKE.KH [Mar94];
- Richardson's number: Ri.1 and Ri.2;
- Endlich's empirical wind index [End64];
- Deformation indices: DEFX$|v|$ and DEFX$|dT/dz|$ [Rea96];
- Wind Shear and Vertical wind shear (VWS);
- Colson–Panovsky index [Col65];

---

[1] The quantity $c_p T + gz$ measured on an isentropic surface, where c_p is the specific heat of air at constant pressure, T is the Kelvin temperature, g is the acceleration of gravity, and z is the height of the isentropic surface. This gives the streamlines for the geostrophic wind when entropy is held constant.

[2] The temperature of an air parcel after dry adiabatic compression or decompression from its actual pressure to 1000 mb.

[3] The level at which a parcel of moist air lifted dry adiabatically would become saturated. It is approximately equivalent to the height of the lowest layer of clouds.

- Brown's index 1 and 2 [Bro73];
- Mountain Wave Turbulence indicator, MWAVE [McC97]; and
- Dutton's empirical index [Dut80].

These indices combine, to different degrees, physical quantities with empirical adjustments. As an example, we review how Ellrod's number is defined. Empirical studies [Ell85] have found that stretching deformation (DST) derived from the u and v wind components by the formula

$$DST = \frac{\delta u}{\delta x} - \frac{\delta v}{\delta y}$$

relates fairly well to the observed CAT, though it produces excessively large threat areas. The value of DST is then combined with the shearing deformation (DSH), computed as

$$DSH = \frac{\delta v}{\delta x} + \frac{\delta u}{\delta y}$$

in the quantity

$$DEF = (DST^2 + DSH^2)^{1/2},$$

which is found to somewhat reduce the extent of the threat areas.

Vertical wind shear (VWS) defined as

$$VWS = \frac{\Delta v}{\Delta z}$$

correlates significantly to CAT as well, and so leads to the product

$$TI1 = VWS \times DEF.$$

TI1 is a simplified version of Ellrod's number. By considering convergence (CVG), defined as

$$CVG = - \left(\frac{\delta u}{\delta x} + \frac{\delta v}{\delta y} \right),$$

the current version of the index has the form

$$TI2 = VWS \times (DEF + CVG).$$

For each index a threshold value is used to provide a dichotomous indicator of CAT potential. These thresholds are set for each index by calibrating the numbers computed by the different algorithms against observed CAT episodes, and this calibration has historically shown a lack of objectivity and consistency across different model outputs and sets of verification data. Our work is focused on rigorously analyzing, discriminating, comparing, and combining the efficacy of the different indices.

3   Data Structure

The verification and training data at our disposal consist of:

- Pilot reports (*pireps*): location, time, and intensity of turbulence episodes encountered along the flight path. A numerical scale labels the intensity of turbulence as light, moderate, or severe in several degrees, so as to favor consistency of the evaluation across the pilots.
- Accelerometer data (*avars*): real-time measurements of all aircraft vertical accelerations along the flight path, whether they are due to turns, climbs, descents, or turbulence encounters.

We use only pireps to infer locations of positive turbulence, and null pireps together with small values of the avars (within $\pm 20\%$ of $1g$) to infer null turbulence locations. To limit the spurious effect of small planes' records, and of turbulence encountered at low altitudes, usually not CAT, we confine our attention to records taken above 20,000 feet.

The observations, especially those based on pilot reports, are limited. They are opportunistic samples, irregularly distributed in space and time and, of course, constrained to the aircrafts' flight paths. The sampling distribution is likely to underrepresent the turbulent areas, because pilots will tend to avoid them when possible (e.g., by following warnings from other pilots who have encountered regions of turbulence earlier in the day). Also, it is expected that data from those pilots used to flying through regions of frequent turbulence may understate the turbulence phenomena and their intensity. In general, whether a turbulence encounter is judged light, moderate, or severe is largely dependent on the past experience of the particular pilot. Furthermore, the severity of the turbulence episode depends on aircraft size. Errors can also be expected based on when and where the pilot chooses to record the episode.

Given these qualifications on pilot reporting, we proceed to match pireps and avars to the CAT indices, as computed by the RUC-60 model. We choose to pair the output of the RUC-60 model with observations taken in a 3 hour window centered at the valid time of the forecast. We match each observation's location to the eight closest grid points, i.e., the vertices of the smallest gridbox which surrounds the data point, and compute the mean value of each index by averaging of the index at those eight points.

The values for pireps are on a raw scale of 0–8. Because of the degree of subjectivity involved in these grades, we use a coarser distinction based on three categories:

- 0 for No or Light turbulence;
- 1 for Moderate turbulence;
- 2 for Severe turbulence.

The case studies presented here are based on datasets pertaining to the following RUC-60 time windows:

- December 5, 1996, 15:00Z and 18:00Z;
- December 11–15, 1997, 15:00Z, 18:00Z, and 21:00Z on each day;

(Z refers to Greenwich Mean Time, and as a reference, 18:00Z is noon EST.)

CAT is mainly a winter phenomenon, whereas convective turbulence (e.g., that due to thunderstorms) dominates the scene in Spring and Summer. Hence we study days in December. Time of day, on the other hand, is not believed to be a crucial component of CAT. The number of observations in each dataset average approximately 500, with a large preponderance of null turbulence observations due to avars. Out of the 6 days under examination, December 5, 1996 was chosen due to the presence of several geographically well-separated areas of turbulence. The set of five consecutive days (December 11–15, 1997) was chosen because of the unusual persistence of an event of moderate to severe turbulence over a large area of the eastern third of the United States.

4   The Single Index Approach

We began by examining the discriminating power of each index separately. This allows comparison to the traditional approach that considers each indicator as a separate piece of information.

One may expect collinearity between indices, since they are computed from the same basic quantities produced by the RUC-60 model. We rule out this concern by visually inspecting scatterplots of their corresponding values, and by inspection of the singular value decomposition of the matrix of indices. Figure 1 shows a subset of paired scatterplots and is indicative of the general behavior of such relations. Neither analyses suggests obvious eliminations, and so we kept the entire suite of 15 indices as potential explanatory variables.

Considering each index separately we studied the distributions of its values, conditional on the different levels of turbulence. Consistent patterns in the quantiles and means of the distributions appear. For example, Figure 2 shows boxplots for six of these indicators. The centers for the conditional distributions shift consistently with the severity of turbulence, but the spread of the distributions is large and overlapping and we can see a number of outliers. This behavior is common to the entire suite of indicators and suggests that each index taken in isolation cannot accurately discriminate different levels of turbulence. It is important to emphasize such inadequacy, because the solutions in place so far have not gone beyond treating these indices as individual indicators, thresholding each of them by a value heuristically determined and generating "threat areas" on the maps on the basis of this binary discrimination.

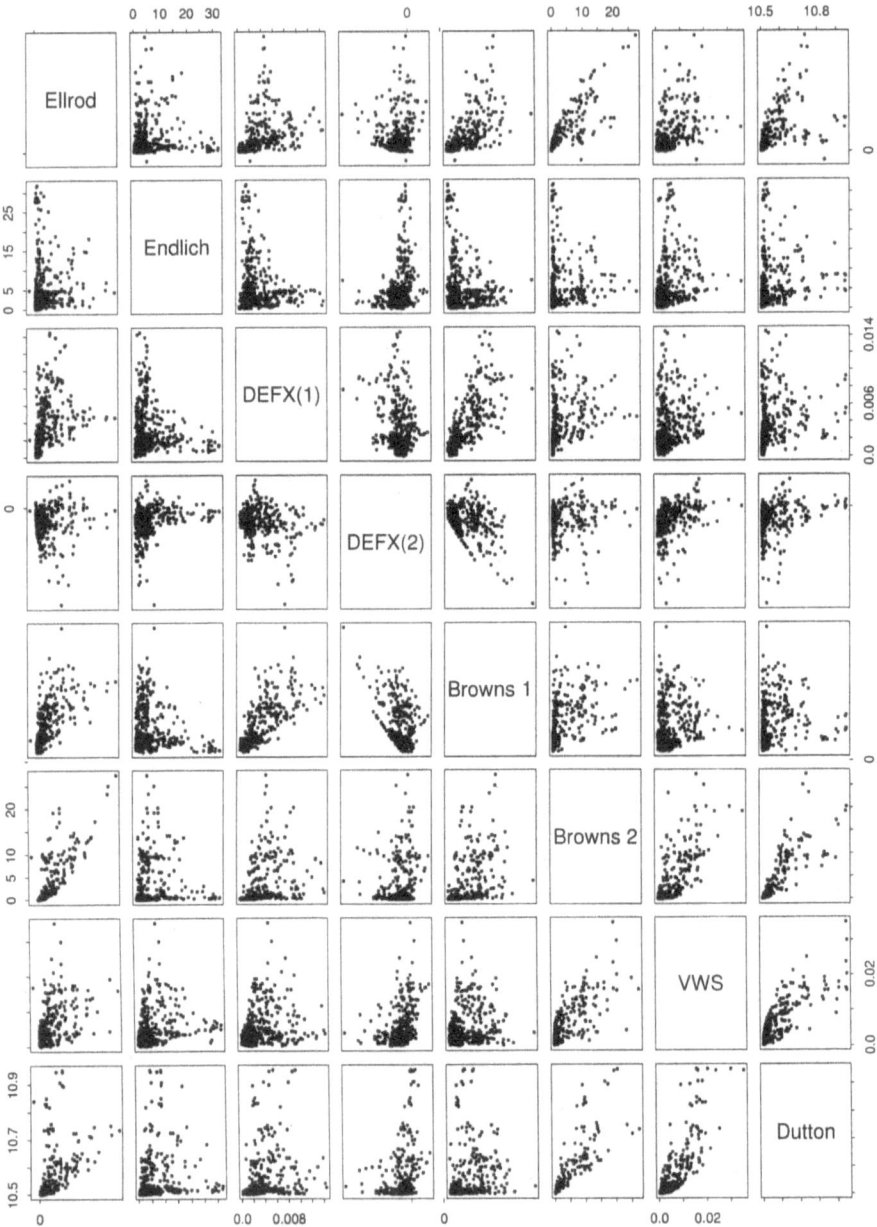

FIGURE 1. Looking for collinearities or more general functional dependence: scatterplots of several pairs of indices' corresponding values.

A standard way of evaluating forecasts is to quantify the trade-off between sensitivity and specificity of the forecast. We define, as a measure of sensitivity, the *probability of detection*, i.e., the ratio of turbulence episodes correctly predicted over the total number of episodes, and as a measure of specificity the *false-alarm rate*, i.e., the ratio of turbulence episodes incorrectly forecast over the total number of episodes forecast. The false-alarm rate is especially critical, in view of the operational value of the forecasts: too large a threat area becomes unmanageable from the pilots' point of view. Also if the false-alarm rate is high, forecasts will tend to be disregarded. When evaluated by these measures we find that thresholding a single index is not feasible for operational forecasts.

5   Modeling Strategy

The objective of our analysis is to find an effective relation between the combined explanatory power of the different indices and the episodes of turbulence, or null turbulence, in the verification data. Considering the quality of the observations, the difference of the spatial scales between the model output and CAT phenomenon, and the difference in temporal scales between the short life of CAT occurrences and the longer range of the forecast, any modeling technique must be flexible to accommodate for nonlinearities and interaction terms in the regression function and adaptive enough to explore the high-dimensional space of the predictors.

We use *multivariate adaptive regression splines* (MARS, [Fri91]) for the regression stage of the analysis, and post-process the results by linear discriminant analysis. The combination of the two is termed *flexible discriminant analysis* (FDA, [Has94]).

5.1   MARS

MARS is a technique for the flexible regression modeling of high-dimensional data. The model takes the form of an expansion in product spline basis functions, where the number of basis functions, as well as the parameters associated with each function (degree and knot locations), are determined adaptively by the data. This procedure is an extension of the recursive partitioning approach to regression (see, for example, the chapter on tree-based models in *Statistical Models in S* [Cla93]) and shares its most useful properties.

5.1.1   Dimension Reduction

It is conceivable that in different subregions of the multidimensional space spanned by the 15 predictors, distinct, lower-dimensional approximations may be sufficient to represent the regression surface. This is the prior expectation in our case, where the single index may be good at discriminating among

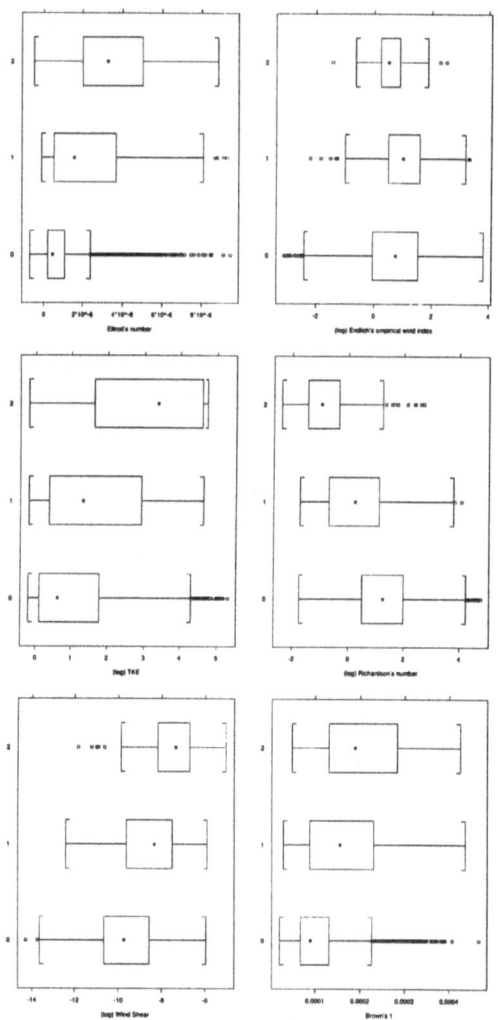

FIGURE 2. Distributions of several indices at the locations of the verification dataset, conditional on the different levels of turbulence observed.

different levels of turbulence when they assume extreme values, but a multivariate classifier may be needed in less extreme cases. MARS is efficient in performing a local variable subset selection. This will be of value in determining redundancies within the set of indices, and in allowing different models to be estimated on the basis of new sets of observations and different temporal and climatic conditions. Besides inheriting these positive attributes from the recursive partitioning approach, MARS improves upon it by fitting functions that are continuous at the boundaries of the subregions, thus enhancing the approximation of the underlying function. MARS also departs from the vertical, hierarchical structure of recursive partitioning to represent functions whose dominant interactions involve only a small fraction of the total number of variables. Where recursive partitioning produces *tree*-like models, MARS comes up with *bush*-like models, with a large number of low-degree interaction terms. This last property is appropriate in this context, because we place low prior belief on interactions that involve a large number of indices at once. Instead, we expect possibly different, small subsets to interact effectively in different regions of the high-dimensional domain spanned by the indices' values.

5.1.2   Functional Form

Within MARS the multivariate spline basis functions take the form:

$$B_m^{(q)}(\mathbf{x}) = \prod_{k=1}^{K_m} [s_{km} \cdot (x_{v(k,m)} - t_{km})]_+^q. \tag{9.1}$$

Here $s_{km} = \pm 1$, q is the degree of the polynomial, and

$$[x]_+ = \begin{cases} 0 & \text{for } x < 0, \\ x & \text{for } x \geq 0. \end{cases} \tag{9.2}$$

K_m is the number of splits that produced B_m. The index $v(k, m)$ identifies the variable involved in the split, t_{km} identifies the knot at which the split takes place.

The result is a model of the form

$$\hat{f}(\mathbf{x}) = a_0 + \sum_{m=1}^{M} a_m \prod_{k=1}^{K_m} [s_{km} \cdot (x_{v(k,m)} - t_{km})]_+^q. \tag{9.3}$$

The model may be recast in a way to highlight the contributions of different degrees of interaction to the global fit; consider the generic expansion:

$$\hat{f}(\mathbf{x}) = a_0 + \sum_{K_m=1} f_i(x_i) + \sum_{K_m=2} f_{ij}(x_i, x_j) \tag{9.4}$$

$$+ \sum_{K_m=3} f_{ijk}(x_i, x_j, x_k) + \cdots$$

In (9.4) the first sum is over all basis functions that involve only a single variable. The second sum is over all basis functions that involve exactly two variables, representing two-variable interactions, and so on. By means of this representation we can identify those variables that enter into the model, whether they enter in isolation or interact with other variables. Dimension reduction occurs when most of the components in this expansion, especially those with many arguments (K_m large), are zero.

5.1.3   Fitting Procedure

The algorithm for fitting this model combines forward fitting and backward pruning. It shares with the recursive partitioning technique, the greedy approach: at each step of the recursive fitting, each of the values of each of the variables available is tested as a possible location for splitting a current region—where a basis function lives—into two adjacent subregions where two new basis functions will be fitted. Either a new, single-term basis function is added to the set of basis functions already in the model, or an interaction term with a pre-existing basis function is chosen, leaving both the parent and child basis function in the model. "Sparing the parents" results in keeping the structure of the final fitted function broader rather than deeper, because in a later step the same parent may generate a new interaction term with a different variable than those involved in its "older" children. This algorithm is how the model selection favors a large number of low-interaction terms rather than a small number of high-interaction terms. As in all recursive techniques, this first stage is followed by the elimination of those terms that fail to provide a significant improvement in the overall goodness-of-fit. The generalized cross-validation criterion of model selection is used in the backward stage, while the residual sum of squares is the quantity minimized at each stage of the forward fitting choices. In all cases, once the form of the model is fixed, the parameters $a_0, \ldots, a_m$ are estimated by least squares.

5.2   FDA

The nonparametric regression estimated by MARS is post-processed by linear discriminant analysis. Only applying the MARS fit to the data yields a scalar score estimated for each observation, and decisions should be made in order to assign the observations to the different categories of turbulence (0, 1, or 2, i.e., No/Light, Moderate, or Severe) on the basis of this score. In fact, the two stages are performed at the same time, by a single S routine called fda[4].

FDA operates by using the multiple response ($J - 1 = 2$ in our case) fit of the MARS model to estimate two optimal and orthogonal sets of scores. The set of observations lying initially in a 15-dimensional space is mapped by the MARS models into a two-dimensional space. In this subspace the discrim-

---

[4]available from   http://lib.stat.cmu.edu/S/mda

ination procedure reduces to the computation of Euclidean distances between each observation and estimated mean values—centroids—for each of the three groups. Under a 0–1 loss function this classification reduces to labeling each observation according to its closest centroid. Under a Gaussian assumption, the distances, along with the prior probabilities of membership, yield a posterior distribution over the three classes of turbulence for each observation.

To explain the FDA approach in more detail, we first review linear discriminant analysis.

5.2.1   Linear Discriminant Analysis and FDA

Linear discriminant analysis (LDA) dates to R.A. Fisher and is a standard tool for classification.

Given a set of N observations y_i, $i = 1, \ldots, N$, each classified into one of J groups γ_j, $j = 1, \ldots, J$, and each associated to a p-dimensional vector of predictors x_i, $i = 1, \ldots, N$, LDA implicitly assumes a multivariate Gaussian distribution for the predictors, with a different mean μ_j, $j = 1, \ldots, J$, but common covariance matrix, Σ. Once $\hat{\mu}_j$ and $\hat{\Sigma}$ are estimated from the observations, the ith observation is classified into the jth category if

$$f(x_i|y_i = \gamma_j) \geq f(x_i|y_i = \gamma_k) \quad \forall k \neq j.$$

Under the Gaussian assumption this is equivalent to finding j so that

$$(x_i - \hat{\mu}_j)'\hat{\Sigma}(x_i - \hat{\mu}_j) < (x_i - \hat{\mu}_k)'\hat{\Sigma}(x_i - \hat{\mu}_k) \quad \forall k \neq j. \tag{9.5}$$

The classification rule in (9.5) can also be formulated as a regression problem. Consider the $(J - 1)$-dimensional hyperplane passing through the J estimated $\hat{\mu}_j$. There exists a matrix U ($p \times J - 1$) which projects each observed x_i, and each estimated $\hat{\mu}_j$ onto this hyperplane, $u = U'x$, $\hat{m} = U'\hat{\mu}$. After this transformation, the u variables have a common identity covariance matrix within groups and the between-group variance has been maximized. That is, the transformation optimally projects the observations so as to make each of the J groups the most homogeneous within itself and best-separated from every other group. The solution to this problem of transforming x into u by an "optimal" linear combination is actually the solution to a multiple linear regression problem. We are regressing the "optimal" $(J-1)$-dimensional scores u_i on the p-dimensional predictors x_i. Also, this two-fold problem of deriving a set of optimal scores and a set of optimal linear combinations of the regressors has been proved equivalent to Canonical Correlation Analysis.

FDA generalizes the discriminant procedure by providing the framework within which more flexible, nonparametric, and adaptive regression techniques are substituted for linear regression. As a result, a set of $J - 1$ nonlinear mappings $h_i(x)$, $i = 1, \ldots J - 1$, provides $J - 1$ scores for each observation. In our case $h(x)$ is the result of applying MARS.

On the basis of these scores LDA acts as a post-processor: the scores are optimized, scaled, and orthogonalized, and provides the new coordinates of each observation in a subspace where Euclidean distances are all that matters. In this lower-dimensional subspace the solution to the discriminant choice can then be worded as the optimal posterior choice, under Gaussian assumption. In the original space of the p predictors, though, the classification rule can be complex and depart from the simple linear approach associated with multivariate normal groups.

5.3   FDA + MARS Algorithm

For details on how the optimal scores are found and the coefficients of the set of regressions estimated, we refer the reader to the original FDA article [Has94] and the corresponding S routines available through the StatLib S archive http://lib.stat.cmu.edu/S/mda.

In schematic terms, FDA + MARS amounts to:

1. An initial choice of a set of $J - 1$ scores for the class identification—the simple indicator matrix whose ij element is 1 if observation i belongs to group j, 0 otherwise is a default choice.
2. A multiresponse regression of the matrix of scores on the predictors by MARS.
3. A reestimation of the scores resulting from Step 2. More precisely, the new scores are the (normalized) eigenvectors of the matrix $\hat{Y}'\hat{Y}$, where $\hat{Y}$ are the $J - 1$ sets of scores predicted by the model at Step 2.

At the end of this procedure each of the observations is associated with $J - 1$ new coordinates, resulting from an optimal (in the sense of most discriminating) mapping of the original p coordinates into the reduced dimensional space.

The final step is to average the coordinates of all observations in each class to estimate the class centroids, in the new lower-dimensional space. Predictions can then be performed on new data points in the form of a posterior probability of membership to each of the J classes, or in the form of a class label. In the $(J - 1)$-dimensional subspace, the posterior probability that observation i belongs to class j is proportional to

$$\pi_j \cdot \exp\{-\frac{1}{2}(u_i - \hat{\mu}_j)'(u_i - \hat{\mu}_j)\}. \tag{9.6}$$

After evaluating the J quantities of the form (9.6) and renormalizing them, we obtain the distribution of posterior probabilities over the J classes and on this basis we can perform the prediction.

6   Implementing FDA + MARS for CAT Forecast

FDA is applied in order to discriminate among the $J = 3$ classes of turbulence observations detailed in Section 3. Therefore, $J - 1 = 2$ sets of optimal scores, together with two sets of regression coefficients (one for each of two MARS models) must be jointly estimated, as explained in Section 5.2.

The default implementation of FDA in S estimates prior probabilities for each category as the corresponding empirical frequencies in the dataset under analysis. This choice is sensitive to the sampling issues detailed in Section 3 and we chose not to adopt it, in favor of a representation less biased toward the category No/Light turbulence.

Hence, two modifications are made:

- Weight each group of observations differently than the empirical frequencies. The best weighting criterion is found to be such that an observation in category 1 weighs 8.5 times an observation in category 0, and an observation in category 2 weighs twice as much; so that, were the representation in the dataset indeed in the exact proportions of $0.85, 0.10, 0.05$, respectively, the sum of the weights in the three groups would be the same.

- Use a noninformative, discrete, uniform distribution as the prior probability for each category.

6.1   Training the Procedure: In-Sample Performances

The FDA + MARS procedure works well, consistently across the number of cases tested (different days and different times of day). Table 9.1 lists the probability of detection of positive turbulence events[5] (pod(y)) and the probability of detection of null turbulence events[6] (pod(n)). $1-\text{pod}(n)$ can be considered as a measure of the false alarm rate. For most of the cases, both values of pod(y) and pod(n) are well above the required lower limit of 0.6.[7]

Figure 3 is a geographic representation of the results for one of the 17 datasets tested. It can be noticed how the main areas of turbulence are correctly identified. It is also evident how the number of No/Light turbulence observations (0's) overwhelms the positive observations, accounting for the tendency of the model to underpredict.

6.2   Testing the Procedure: Prediction Ability

We test the predictive value of our solution in two respects: we first perform a cross-validation exercise by routinely leaving one observation out of the dataset, estimating the model parameters on the remaining data points, and predicting the observation left out by applying the resulting estimates of

---

[5]Computed as the ratio of observations correctly classified as 1 or 2 over the total number of such observations.

[6]Computed as the ratio of observations correctly classified as 0 over the total number of such observations.

[7]In regard to the origin of this lower limit, it is worth pointing out that the concept of "forecast skill" has unresolved issues [Har92]; nevertheless, the weather forecast community has suggested standards that a CAT forecast must meet to be of value [Bro97], [Sha98].

TABLE 9.1. Probability of positive (pod(y)) and null detection (pod(n)) for the 17 models fitted. We highlight those instances which do not meet the standards by an asterisk.

MODEL	pod(n)	pod(y)	obs(n)	obs(y)
12/05/96; 1500Z	0.99	0.90	331	54
12/05/96; 1800Z	0.99	0.67	438	52
12/11/97; 1500Z	0.99	0.73	762	30
12/11/97; 1800Z	0.99	0.54*	759	24
12/11/97; 2100Z	0.99	0.70	681	30
12/12/97; 1500Z	0.98	0.72	644	65
12/12/97; 1800Z	0.99	0.65	692	52
12/12/97; 2100Z	0.99	0.68	701	57
12/13/97; 1500Z	0.98	0.57*	706	42
12/13/97; 1800Z	0.99	0.62	768	43
12/13/97; 2100Z	0.99	0.77	677	31
12/14/97; 1500Z	0.98	0.79	445	33
12/14/97; 1800Z	0.98	0.72	800	58
12/14/97; 2100Z	0.99	0.68	661	50
12/15/97; 1500Z	0.99	0.65	856	17
12/15/97; 1800Z	0.99	0.87	866	16
12/15/97; 2100Z	1	0.69	742	7

the model parameters. In order to make the procedure feasible in an acceptable time span, we fix the form of the MARS function, that is, we fix the form of the terms in (9.3) and let the different subsets of data estimate the value of the parameters a_m. The fixed form is chosen from the result of applying MARS to the entire dataset.

In an effort to produce a prediction scheme as close as possible to the indices' thresholding technique, we rephrase the prediction problem. As detailed in Section 5.2, under a Gaussian assumption, we obtain posterior probabilities of membership in the 0, 1, or 2 category for each observation. At this point, rather than simply assigning the observation to the class having highest posterior probability (a choice largely dependent on the prior distribution used and the form of the loss function implied), we can think of choosing thresholds for these probabilities. In this instance, to simplify the choice of a threshold, we can sum up the probabilities for the two positive classes, and reduce to a binary choice between Null/Light, i.e., 0, and Moderate/Severe, i.e., 1, turbulence (the two categories 1 and 2 are now treated as a single one). As a result of this simplification only one threshold has to be chosen in order to discriminate between high and low posterior probability of CAT. Similarly, the single index approach translates the observed value of the index into a low or high CAT potential on the basis of the threshold chosen for that particular index.

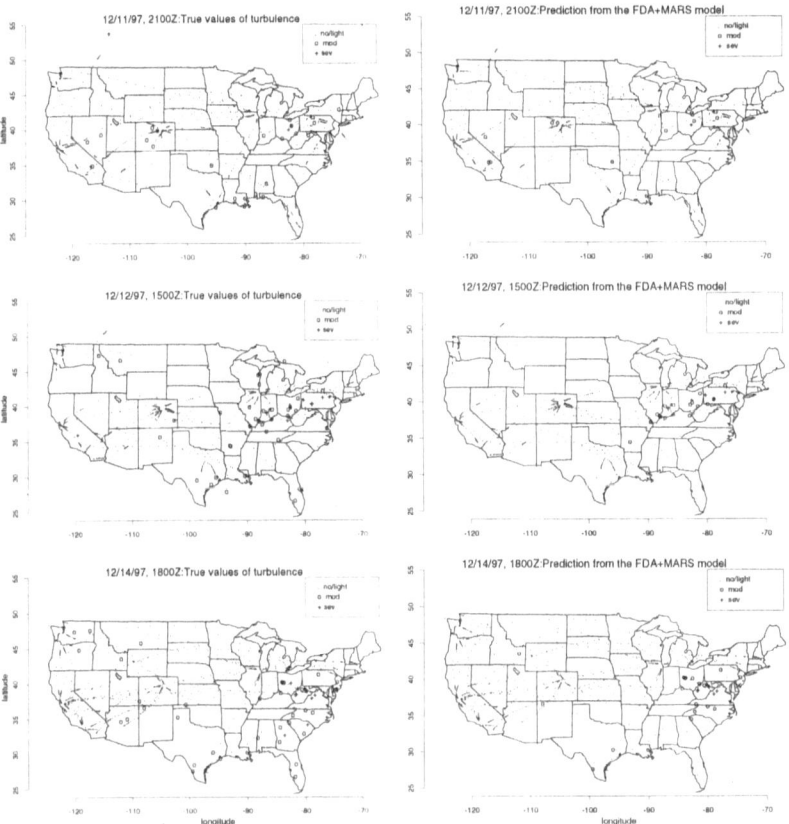

FIGURE 3. A geographic representation of the results.

6.2.1   Cross-Validation Results

We compute the value of this posterior probability by leave-one-out cross-validation. Each data point is in turn left out from its dataset, the model parameters are re-estimated from the remaining points and the posterior probabilities for this omitted observation are calculated based on the binary classification.

We show the results in Figure 4 by lining up boxplots of the posterior probability of "being 1," after separating the observations whose observed level of turbulence is 0 from those whose observed level of turbulence is 1 or 2. The choice of a threshold here corresponds to drawing a horizontal line at a specific value of the probability, and assigning "high posterior probability of turbulence" to the observations above it, and "low posterior probability of turbulence" to the observations below it. We can see that it is possible to choose a value for the threshold to achieve the required standard of 0.6 for both pod(n) and pod(y).

By comparing the first pair of boxplots in each sector to the second pair (corresponding to each dataset's in-sample versus cross-validation results), it is clear that the performance of the discrimination tool is poorer in the cross-validation exercise than in the in-sample outcomes. The boxplots referring to the cross-validation results, in fact, indicate a larger overlap of the two conditional distributions. This can be ascribed to overfitting, an effect often encountered when applying nonparametric adaptive procedures. But in our present setting, it is also the first hint of a problem with data quality. More will be said about this in Section 7.

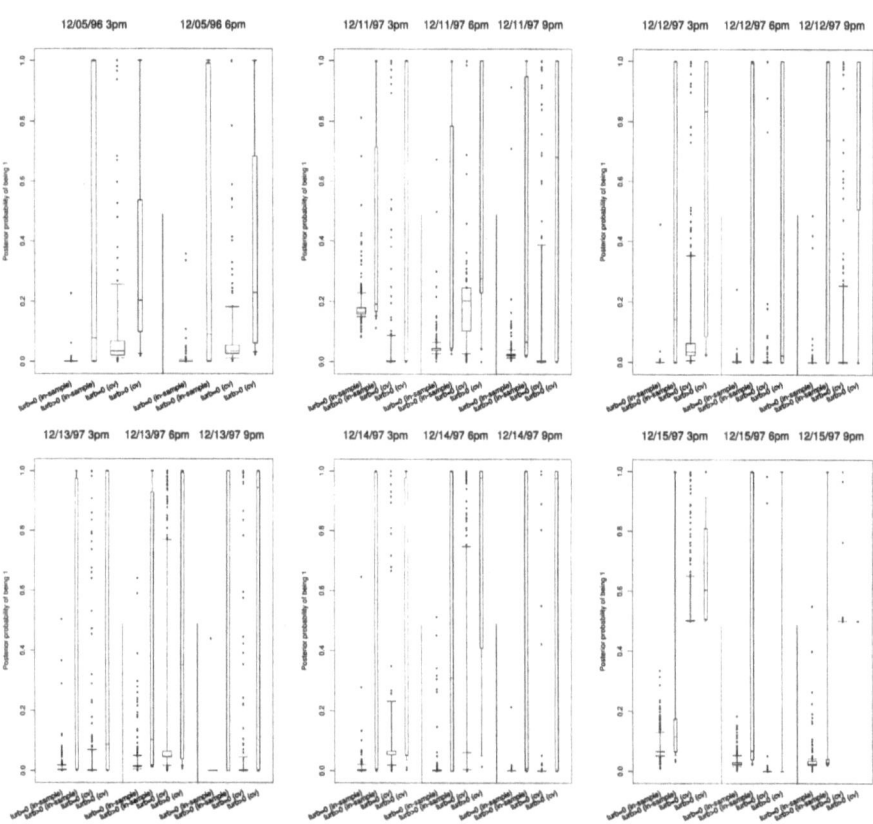

FIGURE 4. Results from cross-validation of the datasets from December 5, 1996, 1500Z to December 15, 1997, 2100Z.

6.2.2   Forecasting CAT

The results from fitting models to past observations are to be used for predicting CAT occurrences ahead in time. This is the real goal of any operational forecast system.

In order to make the number of observations of positive turbulence more substantial, we consolidate several datasets close in time. Experiments with the number of datasets collapsed together showed that care must be taken in order to reach a balance between achieving a rich representation of the recent turbulence environments and maintaining the time window close enough to the current time for prediction. Datasets taken too far back in time represent synoptic or mesoscale conditions that may not hold any longer, and different combinations of indices values would therefore be associated to such different conditions.

We present two cases where a large training set is used to estimate a binary classification model, and the posterior probability of turbulence is derived for a set of out-of-sample observations. The focus of this exercise is the comparison with the single indices' discrimination abilities. We choose six of the most commonly used indices (TKE, Richardson's number, Ellrod's, Dutton's, Brown's 1, Shear) and use a threshold on each index as a predictor of turbulence. For any threshold value and index one can obtain $pod(y)$ and $pod(n)$ as a measure of performance. By running the threshold over all the observed values of the index we produce a curve in the $pod(y)$–$pod(n)$ coordinates. This curve is a comprehensive summary of the performance of the index. We can choose a point on the curve (i.e., a $pod(y)$, $pod(n)$ pair) by fixing the threshold at the corresponding level. But we cannot improve on both $pod(y)$ and $pod(n)$ simultaneously. With the single indices' result we include the results of the FDA method, applying a threshold to the posterior probability.

Figure 5 shows two such results. The curves trace the $pod(y)$–$pod(n)$ trade-off, and show that the multidimensional models estimated by the FDA + MARS technique have better performance with respect to the single indices' solutions. The thick solid lines, corresponding to the results obtained by FDA + MARS dominates the results obtainable by any of the single indices used in isolation.

7   MARS as Variable Subset Selection

A scientific goal of this chapter is to highlight how the CAT indices, considered in isolation, are inferior to a multidimensional approach that combines indices in a statistical model. By considering the form of the functions fitted by MARS, we can further substantiate this claim.

MARS, as an adaptive technique, selects the variables entering the final model equation, and the position of the knots for each variable. In Table 9.2 we list the 15 models fitted to the period December 11–15, 1997, with respect to the number of times (i.e., in how many terms) each index appears in each model. (Here we do not distinguish whether it enters the model in isolation or in interaction terms, combined with other indices.)

As Table 9.2 shows, none of the indices have been left out but some of them,

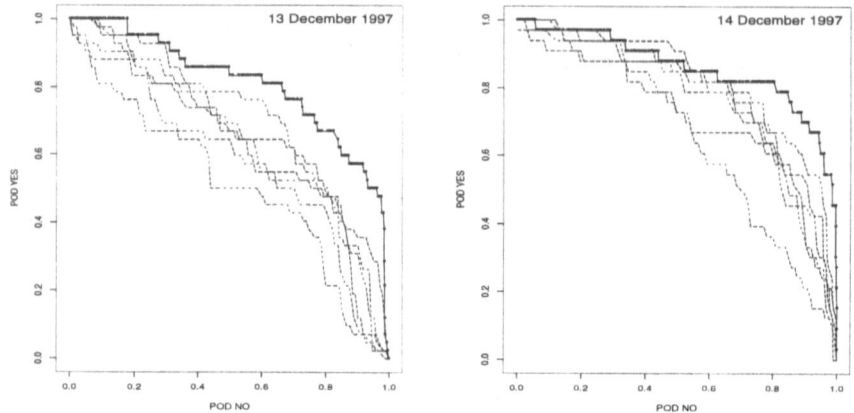

FIGURE 5. Predicting 13 December 1997, 1500Z by all previous datasets; predicting 14 December 1997, 1500Z by 11, 12, 13 December 1997, 1500Z; solid line with points overlayed: FDA + MARS model; dotted lines: single indices' thresholding.

TABLE 9.2. The usage of the different indices in 15 models. The number at the intersection of column i and row j indicates the number of terms in model i which include index j.

day	12.11.97			12.12.97			12.13.97			12.14.97			12.15.97		
index	15 Z	18 Z	21 Z	15 Z	18 Z	21 Z	15 Z	18 Z	21 Z	15 Z	18 Z	21 Z	15 Z	18 Z	21 Z
Ell	0	14	6	1	1	0	2	1	0	1	1	1	8	2	0
TKE	0	0	2	0	1	6	0	2	0	0	0	0	0	12	0
R1	1	0	2	4	0	1	2	3	0	1	0	3	0	1	3
TKEKH	2	0	1	0	0	0	0	1	1	0	0	0	0	0	0
End	14	3	0	0	0	0	1	1	1	1	0	2	0	1	0
DEF1	1	5	1	1	1	0	9	1	2	2	3	1	2	2	2
DEF2	0	0	0	1	1	2	0	5	1	0	0	1	5	0	0
Shear	1	1	0	1	4	0	0	0	0	0	0	0	0	1	0
CoPa	1	3	14	5	4	6	2	4	4	4	5	5	1	5	0
B1	1	0	1	2	0	1	12	0	4	0	1	1	1	2	1
B2	0	0	0	13	4	14	0	1	4	1	2	5	10	0	1
R2	2	0	2	0	0	1	2	1	2	0	0	1	1	0	3
VWS	2	1	7	0	3	3	2	0	3	0	2	1	0	0	0
Dutt	1	5	0	3	1	0	0	0	0	0	2	0	0	0	1

like TKE.KH, DEFX$|dT/dz|$, Richardson's numbers, and Dutton rarely appear. Ellrod's, Colson–Panovski, and Brown's 2 are heavily represented across almost all models, in single terms or in two- and three-degree interaction terms. The important idea from this table, however, is how different the models appear to be, even those from adjacent time windows. In fact, the 5 days under examination showed a persistent episode of mid- to upper-level severe turbulence over the eastern third of the United States, and the general conditions of the atmosphere were stable throughout the entire period. Although this would suggest a consistency in the form of the models fitted, it does not appear to be the case. We infer that the adaptive skills of MARS are coping with the inconsistencies in the data quality. This flexibility and adaptivity can also explain the poorer performance in the out-of-sample and cross-validation exercises, if compared to the in-sample discrimination ability.

TABLE 9.3. The interactions between the different indices in the 15 models. The number under the ith column of the jth row indicates the number of terms in the 15 models which include both variable i and variable j as an interaction effect. The table is by construction symmetric. The main diagonal contains the total number of terms in the 15 models that use the corresponding index, alone or in interaction terms, i.e., the sum of the rows of Table 9.2

index	Ell	TKE	R1	TKEKH	End	DEF1	DEF2	Shear	CoPa	B1	B2	R2	VWS	Dutt
Ell	38	3	2	0	3	11	5	1	11	4	1	3	2	5
TKE	3	23	3	1	3	2	1	1	7	3	4	0	2	0
R1	2	3	21	1	1	2	3	0	2	3	6	3	0	2
TKEKH	0	1	1	5	2	0	1	0	1	0	0	0	1	0
End	3	3	1	2	23	2	0	0	5	1	3	1	1	0
DEF1	11	2	2	0	2	33	2	2	12	10	4	2	4	0
DEF2	5	1	3	1	0	2	16	1	2	2	2	2	2	0
Shear	1	1	0	0	0	2	1	8	0	1	1	1	2	0
CoPa	11	7	2	1	5	12	2	0	63	5	7	6	7	0
B1	4	3	3	0	1	10	2	1	5	27	1	3	2	0
B2	1	4	6	0	3	4	2	1	7	1	44	1	5	4
R2	3	0	3	0	1	2	2	1	6	3	1	15	0	0
VWS	2	2	0	1	1	4	2	2	7	2	5	0	24	1
Dutt	5	0	2	0	0	0	0	0	0	0	4	0	1	13

The large number of two- and three-degree interaction terms are detailed in Table 9.3. Each cell represents an interaction between two indices, and the number in the cell indicates the number of terms that use that interaction, out of all the 15 models. This reinforces the conclusion that the multidimensional approach to the problem adds valuable information to the predictive process.

8 Conclusions

The results presented in these case studies are far from conclusive, and further work must be devoted to fitting better models using better quality verification data and more effective indices. Several specific directions are:

- Given that training the model requires large amounts of data, one should build training sets, limiting the huge amount of 0 observations, and enhancing the representation of positive instances of CAT encounters. Balanced sampling from the available past observations could be considered for this purpose.
- The time scale for matching observations to the RUC-60 model output may warrant further exploration. Narrower time windows centered at the valid time of the weather forecast might improve the prediction. On the other hand, larger windows could accommodate recording errors in the pireps data, and cope with their sparseness.
- Subsetting the data available by geographic area and altitude, and fitting regional models, may lead to more stable variable subset selection and, hence, improved forecasts.
- Using a large, historical set of observations able to provide a reliable climatology of turbulence, may represent a valuable piece of additional information, from which more informed prior probabilities may be drawn.

Spatial Structure of the SeaWiFS Ocean Color Data for the North Atlantic Ocean

Montserrat Fuentes
North Carolina State University, Raleigh,
NC 27965, USA

Scott C. Doney
National Center for Atmospheric Research, Boulder,
CO 80307, USA

David M. Glover and Scott J. McCue
Woods Hole Oceanographic Institution, Woods Hole,
MA 02543, USA

1 Introduction

As part of a longer-term marine ecosystem modeling project, we examine the spatial structure of satellite-derived ocean color data for the North Atlantic Ocean. Ocean color is considered a proxy for surface layer phytoplankton chlorophyll concentrations, and the large-scale ocean color field is governed by the seasonal distributions of light, nutrients, and upper ocean mixing (e.g., [Wro89]). On the so-called mesoscale (approx. 10–200 km), ocean color variability is modulated by biological sources and sinks (e.g., phytoplankton growth, zooplankton grazing), ocean flow, and mixing [Ste78], [Hau78]. The spatial correlation function of ocean color provides a useful measure for quantifying these biological–physical interactions [Yod87], [Was98], [Abb98] and discriminating among theoretical models. It is also a necessary component of future work to objectively analyze ocean color images (e.g., [Dav86]).

In this chapter, the spatial scales of satellite chlorophyll images are determined using the semivariogram from geostatistics (e.g., [Cre93]). The local correlation patterns derived from the semivariograms are both anisotropic (varying with direction) and nonstationary (varying in space and time). The resulting zonal (east–west) and meridional (north–south) distributions of chlorophyll variance and ranges of spatial correlation are then compared to physical, mesoscale properties for the North Atlantic Ocean such as the Rossby deformation radius and the eddy kinetic energy. The SeaWiFS ocean color data used in this study are described in Section 2, and the necessary geostatistical tools, including the robust empirical semivariogram and semivariogram models, are reviewed in Section 3. In Section 4, we discuss in detail example calculations of the semivariogram and model parameters for a single $5° \times 5°$ box in the middle of the North Atlantic. The North Atlantic Ocean correlation patterns are presented and analyzed in Section 5, followed by conclusions and some final remarks in Section 6.

2   SeaWiFS Ocean Color Data

Phytoplankton are small, generally single-celled organisms that grow in the sea by converting sunlight into chemical energy by the process of photosynthesis. Chlorophyll, one of the main light-harvesting pigments within phytoplankton, absorbs sunlight in the blue and red spectral ranges. The amount or concentration of chlorophyll in seawater is a primary factor in determining the penetration depth and spectral quality of light in the upper ocean. Ocean color instruments such as the Sea-viewing Wide Field of view Sensor (SeaWiFS, [Hoo93]) rely on this change in the spectral composition of radiance to quantify the chlorophyll concentration (mg m^{-3}). The SeaWiFS instrument measures eight spectral bands in the visible and near-infrared. Concentrations of chlorophyll are calculated from ratios of radiances of different bands [O'R98]. Only a small fraction of the light observed by the satellite has upwelled from below the ocean surface, the remainder being reflected sunlight from the surface, atmospheric gases, or aerosols. In addition to the required atmospheric corrections, the SeaWiFS data are also processed to remove pixels with land or clouds before geophysical variables such as chlorophyll are computed.

The SeaWiFS instrument became operational in September 1997. Its Sun-synchronous polar orbit (altitude 705 km, orbit period approximately 100 minutes, image swath width roughly 1500 km) provides near-global coverage every 2 days with a nominal resolution of about 1 km when the sensor is directly overhead. The SeaWiFS data are reported at different processing levels: from raw spacecraft telemetry (Level 0) to geophysical variables on standard grids (Level 3) and at different spatial resolutions [McC98]. The high-resolution Local Area Coverage (LAC), with nominal resolution of 1 km $\times$ 1 km, is broadcast continually by the spacecraft. The Global Area Coverage (GAC), with a nominal resolution of 4 km $\times$ 4 km, is a continuous, global, subsampled version of the LAC and is stored aboard the spacecraft for later transmission to ground stations. A daily, (Level 3)-resolution product is produced by averaging all of the GAC data from a single day after binning the data to an equal-area grid (81 km^2). This binned grid is difficult to work with because the number of longitudinal grid cells varies with latitude. Thus, the primary data used in this study are projections of the daily-binned product onto a global, equal-angle grid ($2\pi/2048$) with a nominal resolution of 9 km $\times$ 9 km, referred to as standard mapped images. When higher resolution is required, we use chlorophyll concentration data from the Bedford Institute of Oceanography, Nova Scotia station, processed to Level 2, still in the form of satellite images (i.e., with geometric distortions).

The geostatistical measures discussed in the next section give a complete description of the spatial distribution of Gaussian, stationary processes. Unfortunately, neither of these two assumptions hold completely for ocean chlorophyll. Spatial trends, in particular, can produce spurious effects in the semivariogram [Cla79], [Cre93]. Two steps are taken to normalize the chlorophyll data prior to analysis. First, the natural log transformation ln(Chl) is applied to the data following the arguments of Campbell [Cam95] and others, that oceanic bio-optical variability is approximately log-normal. Second, the large-scale spatial patterns in chlorophyll were removed by subtracting a smoothed version of the monthly average from the daily chlorophyll, resulting in *daily anomalies*. Specifically, monthly means for each pixel were calculated on the log scale and then the resulting image was smoothed using a two-dimensional kernel with a 0.5° bandwidth. The resulting daily anomalies are merged into two-day blocks, the time it takes the sensor to completely sample the globe. Due to the presence of clouds and other ocean–atmosphere effects (e.g., aerosols, Sun-glint), the number of samples that go into the composite for any particular pixel can vary widely, and thus the composite is a biased estimator of the mean field. In this chapter we work with the daily chlorophyll concentrations, therefore the bias in the composite should not affect our results significantly.

3   Semivariograms and Other Tools in Spatial Statistics

Geophysical data typically exhibit variation across a wide range of spatial scales, that often may be explained as a combination of large-scale trends and small-scale variability. A *regionalized variable* Z, such as the chlorophyll concentration, is defined in geostatistics [Jou78] as

$$Z(\mathbf{x}) = \overline{Z}(\mathbf{x}) + \epsilon(\mathbf{x}), \qquad (10.1)$$

where $\overline{Z}(\mathbf{x})$ is the large-scale field and $\epsilon(\mathbf{x})$ is local variation. As a statistical model we will treat $\epsilon(\mathbf{x})$ as a random variable. The variation in $\epsilon(\mathbf{x})$ includes both measurement noise and small-scale spatial processes. A major objective of geostatistics is to characterize the covariance function of $\epsilon(\mathbf{x})$. This is also a necessary ingredient for spatial prediction procedures such as kriging.

The *semivariogram* measures the local spatial variation of a random field by describing how sample data are related with distance and direction. It is the crucial concept in geostatistics [Jou78]. In general, two closely neighboring points are more likely to have similar values than two points farther apart.

The semivariogram function γ is defined as

$$\gamma(\mathbf{v}) = \frac{1}{2}\text{var}\{Z(\mathbf{x} + \mathbf{v}) - Z(\mathbf{x})\}, \qquad (10.2)$$

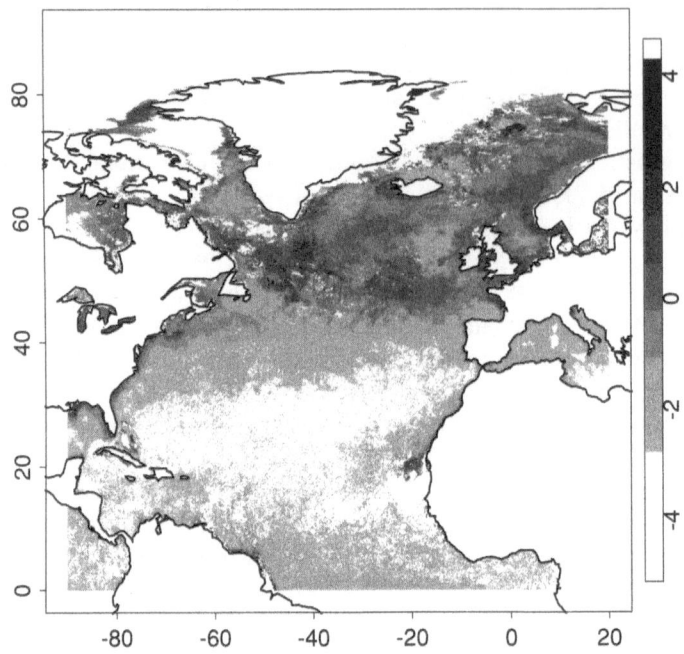

FIGURE 1. Natural log of the composite SeaWiFS chlorophyll concentration (mg m^{-3}) for June 1998 over the North Atlantic Ocean.

where **v** is the vector distance (considering direction) separating two points in space. When the covariance function, C, of the process Z exists, the semivariogram can also be written as

$$\gamma(\mathbf{v}) = C(\mathbf{0}) - C(\mathbf{v}) \qquad (10.3)$$

and so the two functions are closely interrelated. The observational or empirical semivariogram can be efficiently computed for data on a regular grid from the spectral density function (the Fourier transform of the covariance) using the Fast Fourier Transformation (FFT) and smoothed periodogram. Unfortunately, geophysical datasets are often characterized by unequally spaced sampling and/or incomplete grids. Two related methods for estimating the empirical semivariogram are discussed in Subsection 3.2.

The semivariogram of a process Z is a function of the second-order moments, and for Gaussian fields, which by definition depend only on the mean and the variance, the semivariogram uniquely defines the underlying process. For non-Gaussian fields, the semivariogram can still be used as a measure of the local variation.

3.1 Stationarity and Isotropy

Consider a spatial process $\{Z(\mathbf{x}), \mathbf{x} \in R^2\}$. The process Z is said to be *strictly stationary* if the joint distribution of

$$(Z(\mathbf{x}_1), Z(\mathbf{x}_2), \ldots, Z(\mathbf{x}_m))$$

is the same as

$$(Z(\mathbf{x}_1 + \mathbf{v}), Z(\mathbf{x}_2 + \mathbf{v}), \ldots, Z(\mathbf{x}_m + \mathbf{v}))$$

for any m spatial points $\mathbf{x}_1, \ldots, \mathbf{x}_m$ and any $\mathbf{v} \in R^2$.

The process Z is said to be *second-order stationary*, or *weakly stationary*, if the mean is constant and the covariance function satisfies

$$\operatorname{cov}\{Z(\mathbf{x}_1), Z(\mathbf{x}_2)\} = C(\mathbf{x}_1 - \mathbf{x}_2) \qquad \text{for all} \quad \mathbf{x}_1, \mathbf{x}_2 \in R^2.$$

A strictly stationary process, assuming all variances are finite, is also second-order stationary. The converse is false in general. However, a *Gaussian* process that is second-order stationary is also strictly stationary. Throughout this chapter we work with Gaussian processes, therefore there is no need to differentiate between the two types of stationarity.

Suppose γ is the semivariogram function for Z. If $\gamma(\mathbf{v}) = \gamma_0(|\mathbf{v}|)$ for some function γ_0, then the process is called *isotropic*.

3.2 Estimating the Empirical Semivariogram.

The traditional semivariogram estimate $\hat{\gamma}$ suggested by Matheron [Mat71] is

$$\hat{\gamma}(\mathbf{v}) = \frac{1}{2N(\mathbf{v})} \sum_{N(\mathbf{v})} (Z(\mathbf{x}_i) - Z(\mathbf{x}_j))^2, \tag{10.4}$$

$\mathbf{x}_1, \ldots, \mathbf{x}_m$ where $N(\mathbf{v})$ are the number of data locations $\mathbf{x}_i$ and $\mathbf{x}_j$ separated by $\mathbf{v}$. Sometimes, it is also desirable to consider the semivariogram as a function of the length of $\mathbf{v}$, $h = \|\mathbf{v}\|$, independent of direction, and we will then write $\hat{\gamma}(h)$. The empirical values of $\hat{\gamma}(\mathbf{v})$ at different separations $\mathbf{v}$ are correlated, a point to be considered when fitting model semivariogram curves using methods such as nonlinear least squares that assume independent observations.

Cressie and Hawkins [Cre80] propose an alternative, robust (to contamination by outliers), estimate of the semivariogram [Cre93, p. 75]:

$$\bar{\gamma}(\mathbf{v}) = \frac{1}{2} \left(\frac{1}{N(\mathbf{v})} \sum_{N(\mathbf{v})} |Z(\mathbf{x}_i) - Z(\mathbf{x}_j)|^{1/2} \right)^4 \bigg/ \left(0.457 + \frac{0.494}{N(\mathbf{v})} \right). \tag{10.5}$$

The reasoning behind (10.5) is that for Gaussian data $Z(\mathbf{x})$ with constant mean, $(Z(\mathbf{x}_i) - Z(\mathbf{x}_j))^2$ is a chi-squared random variable on one degree of freedom. The power transformation that makes a chi-squared random variable most Gaussian-like is the square root. Thus, various location estimators can be applied to the quantity $|Z(\mathbf{x}_i) - Z(\mathbf{x}_j)|^{1/2}$ which, when normalized for bias, yield robust semivariogram estimators. The normalizing constant $\{0.457 + 0.494/N(\mathbf{v})\}^{-1}$ makes $\bar{\gamma}(\mathbf{v})$ an unbiased estimate of $\gamma(\mathbf{v})$.

There are two practical rules [Jou78] that should be considered when estimating an empirical semivariogram:

- The empirical semivariogram should only be considered for distances h for which the number of pairs is greater than 30.
- The *distance of reliability* for an experimental semivariogram is $h < D/2$, where D is the maximum distance over the field of data.

3.3 Parameters of the Semivariogram: Nugget, Sill, and Range

A main goal of geostatistical modeling is to construct a semivariogram model that best estimates the spatial correlation of the observed field. Most semivariograms are defined through several parameters; namely, the *nugget*, *sill*, and *range*:

- *nugget:* represents unresolved, microscale variation and/or measurement error. It is estimated from the empirical semivariogram as the value of $\gamma(h)$ as $h \to 0$;
- *sill:* the value of $\gamma(h)$ for $h \to \infty$ representing the variance σ^2 of the random field; and
- *range:* a scale parameter controlling the degree of correlation with distance $\mathbf{v}$ (the precise definition depends on the form of the semivariogram model).

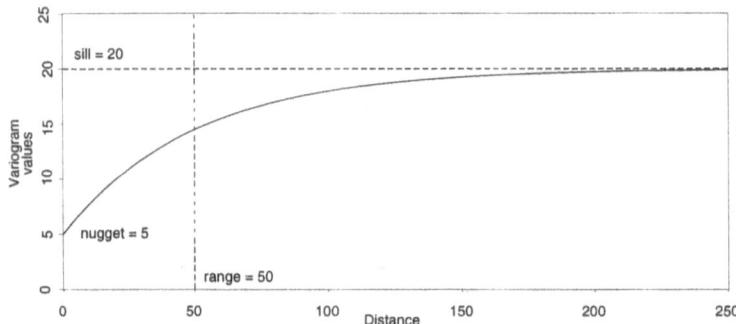

FIGURE 2. A hypothetical exponential semivariogram showing the sill and range parameters as well as a nugget effect. For an exponential model, the range is the distance at which the spatial autocorrelation drops to $1/e$ of its value at zero lag (in the absence of the nugget effect).

3.4   Some Isotropic Semivariogram Models

Journel and Huijbregts [Jou78] present several parametric semivariogram models. Here we just highlight the exponential, spherical, and linear.

- *Exponential model*

$$\gamma(h) = \begin{cases} 0 & \text{if } h = 0, \\ c_0 + (\sigma^2 - c_0)\{1 - \exp(-h/r)\} & \text{if } h \neq 0. \end{cases} \qquad (10.6)$$

- *Spherical model*

$$\gamma(h) = \begin{cases} 0 & \text{if } h = 0, \\ c_0 + (\sigma^2 - c_0)\left\{\frac{3}{2}(h/r) - \frac{1}{2}(h/r)^3\right\} & \text{if } 0 < h \leq r, \\ \sigma^2 & \text{if } h \geq r. \end{cases} \qquad (10.7)$$

- *Linear model*

$$\gamma(h) = \begin{cases} 0 & \text{if } h = 0, \\ c_0 + bh & \text{if } h \neq 0. \end{cases} \qquad (10.8)$$

In the spherical and exponential models, the parameter c_0 is the nugget ($c_0 \geq 0$), σ^2 denotes the sill (or variance of the process), and r is the range ($r \geq 0$). In the exponential case, the range is the distance at which the spatial autocorrelation drops to $1/e$ of its value at zero lag (in the absence of the nugget effect). In the linear model, the only two parameters are the nugget, c_0, and the slope, b. Figure 3 shows an example of an exponential, a linear, and a spherical semivariogram.

For spatial data, the exponential model is generally appropriate and commonly used. The linear model is useful when the process under consideration has unbounded variance, an example of a case where the semivariogram can be defined but the covariance function $C(\mathbf{v})$ cannot. Here we work with exponential models. Another valid semivariogram model that is frequently used for spatial processes is [Mat86]:

- *Matern class*

$$\gamma(h) = \begin{cases} 0 & \text{if } h = 0, \\ c_0 + (\sigma^2 - c_0)\left\{1 - \frac{(2h\sqrt{\nu}/r)^\nu K_v\left(2h\sqrt{\nu}/r\right)}{2^{\nu-1}\Gamma(\nu)}\right\} & \text{if } h \neq 0, \end{cases} \qquad (10.9)$$

where the parameters ν and r are greater than zero, the parameter ν is the degree of smoothness of the process, and K_ν is the modified Bessel function of the second kind. Whittle [Whi54] proposed the case $\nu = 1$ as the most appropriate for spatial processes.

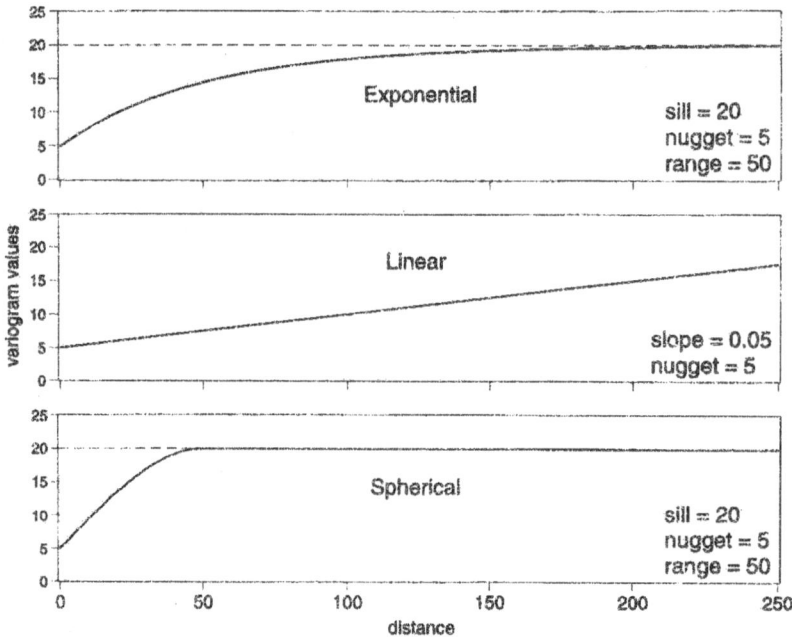

FIGURE 3. Example exponential, linear and spherical semivariograms.

4 Ocean Color Semivariograms

4.1 Spatial Analysis for the Ocean Chlorophyll

Semivariograms computed from the SeaWiFS Level 3 ocean color data for June 1998 are presented in Figure 4. The data are from a 5° × 5° box (470 km × 557 km) located at 30°W–35°W and 30°N–35°N in the open ocean, North Atlantic, in a region of relatively low chlorophyll concentrations (Figure 1). The size of this box was chosen after some experimentation to minimize the spatial heterogeneity of the ocean color data while retaining sufficient spatial structure to resolve the full span of the mesoscale (10 km–200 km) features. Separate semivariograms are estimated for the zonal (east–west) and meridional (north–south) directions as well as for the two diagonals (southwest–northeast and northwest–southeast), shown as different panels in Figure 4, to highlight any potential directional anisotropy. For a single month, the estimated semivariogram values versus distance (in km) are computed using the robust semivariogram form (10.5) for 15 pairs of daily chlorophyll anomalies (we combine consecutive days to improve data coverage as explained in Section 2). The approximate bin size (nominally 9 km) differs with direction and latitude.

The boxplots in Figure 4 represent the resulting distribution for the 15 two-day pairs. Although there is considerable scatter in the individual two-

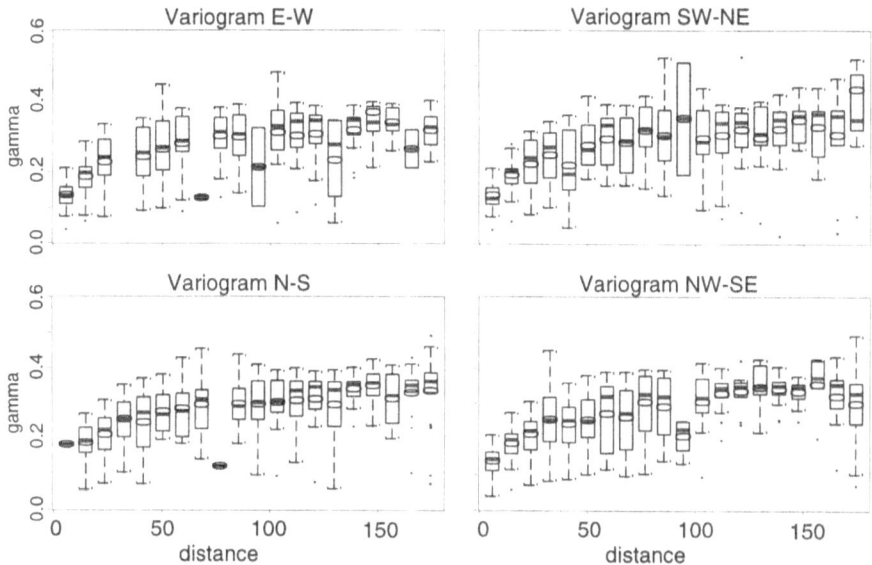

FIGURE 4. Semivariogram for the log chlorophyll anomalies for June 1998, in a region bounded by 30°W–35°W and 30°N–35°N in the middle of the North Atlantic Ocean. Each panel shows the estimated semivariogram in a different direction, and each boxplot represents the distribution of the semivariogram values for the anomalies at a fixed distance. The median value in each boxplot is shown as a line across the box, and the mean as a circle.

day pair semivariogram estimates at any particular distance, the mean and median values follow well-defined curves with $\bar{\gamma}(\mathbf{v})$ increasing from about 0.15 near the origin to a uniform sill of about 0.35 beyond 100 km.

Quantitative estimates of the range, the sill and the nugget are derived by fitting the exponential model form (10.6) to the median value for each boxplot (Table 4.1). A Gauss–Newton nonlinear least squares algorithm is used to find the set of model parameters that minimizes the sum of the squared residuals (SSR) between the response and the prediction. Approximate standard errors are estimated by linearizing the semivariogram model about the final convergence point and then applying traditional linear regression where the estimated standard deviations of the estimates are functions of the SSR.

The ocean color data for the 30°W–35°W and 30°N–35°N box (Figure 4) show a weak directional anisotropy (see Table 4.1), with a larger estimated range parameter in the meridional (north–south) direction (69 km) relative to either the zonal or diagonal directions. The gradient in the large-scale chlorophyll field for this region is also oriented approximately north–south, and the larger range calculated in that direction may reflect a partial aliasing of the nonstationarity in the mean, despite the removal of the smoothed monthly

TABLE 10.1. Model fits of the semivariogram parameters for the ocean chlorophyll anomalies for a $5° \times 5°$ box in the North Atlantic using different directions, for the month of June 1998. The table shows 95% confidence intervals for the nugget, sill, and range.

Longitude	Latitude	Direction	Nugget	Partial sill	Range	Residual	n
30°W–35°W	30°N–35°N	N–S	0.15 ± 0.04	0.21 ± 0.02	69 ± 32	0.014	16
30°W–35°W	30°N–35°N	NW–SE	0.13 ± 0.06	0.23 ± 0.04	63 ± 48	0.028	16
30°W–35°W	30°N–35°N	E–W	0.09 ± 0.12	0.22 ± 0.10	32 ± 26	0.034	15
30°W–35°W	30°N–35°N	SW–NE	0.11 ± 0.06	0.24 ± 0.06	47 ± 28	0.028	16

field. The total sill parameter (partial sill plus nugget) is approximately the same in each direction, roughly 0.35, and the nugget effect is larger in the north–south direction, with a value of 0.15 versus around 0.11 in the other directions. The nugget, however, is not well defined from the 9 km Level 3 data and is quite sensitive to the fit through one or two data points near the origin. The observed weak anisotropy is expected for an open ocean site away from strong boundary currents. There is more clear evidence of anisotropy in the coastal regions (not shown).

4.2 Semivariograms in the Original Scale

As noted in Section 2, the ocean color data is log-transformed prior to the calculation of semivariograms based on the assumption that ocean bio-optical properties are roughly log-normal [Cam95]. This assumption can be tested, to first order, for the SeaWiFS data by seeing whether the observed variances are proportional to the squared average concentrations across a range of mean chlorophyll concentrations. Figure 5 compares the squared mean chlorophyll concentration versus the variance (both on log scales) for a set of small sub-regions ($5° \times 5°$) along a line of constant longitude 30°W–35°W from 0°N to 90°N. The relationship is roughly linear, supporting our use of the log transform for these data. We could alternatively use a *relative variogram* on the raw data, see [Cre85], which is roughly equivalent to estimating variograms on the appropriately transformed data. Clearly, the core methodology in spatial statistics is built on Gaussian fields and so, whenever possible, it is best to transform the data to an approximate normal distribution prior to calculating the semivariogram. As discussed below, the semivariogram sill and nugget can then be interpreted in the original scale as needed. In general, the range parameter is approximately the same in both scales.

The transformed ocean color anomalies $A_{\mathbf{x}t}$, at a location $\mathbf{x}$ and time t, are computed by taking the logarithm of the original data $Z_{\mathbf{x}t}$ and then subtracting the time-mean composite $\overline{\log(Z_{\mathbf{x}})}$:

$$A_{\mathbf{x}t} = \log(Z_{\mathbf{x}t}) - \overline{\log(Z_{\mathbf{x}})}. \tag{10.10}$$

If we consider a small enough region in space, we can write the time-mean composite, $\overline{\log(Z_{\mathbf{x}})}$, independently of $\mathbf{x}$, as a constant $\overline{\log(Z)}$ for the region

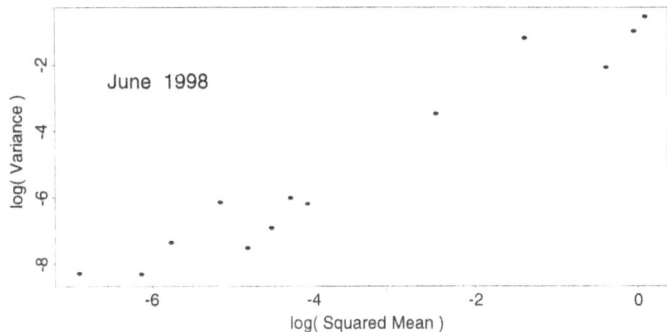

FIGURE 5. Comparison of the squared mean chlorophyll concentration and variance on log scales in small $5° \times 5°$ subregions for June 6, 1998. The longitude of the 18 subregions is constant $30°W–35°W$ and the latitude varies from $0°N$ to $90°N$.

The sill and the nugget on the original scale then are simply $\exp[(2\ \overline{\log(Z)})]$ times the exponential of their values for the transformed anomalies.

Define $\gamma_a(\mathbf{h})$, the semivariogram for the concentration of chlorophyll as measured by the transformed anomalies, $A_{\mathbf{x}t}$, at a distance $\mathbf{h}$, and σ_a^2 the sill for the process $A_{\mathbf{x}t}$, then we have [Mej74]:

$$\gamma_e(\mathbf{h}) = \exp(\sigma_a^2\, 2) - \exp(\sigma_a^2\, 2 - \gamma_a(\mathbf{h})), \tag{10.11}$$

where $\gamma_e(\mathbf{h})$ is the semivariogram at a distance $\mathbf{h}$ for $e^{A_{\mathbf{x}t}}$. We could use the previous expression to obtain the nugget effect for $e^{A_{\mathbf{x}t}}$. The relationship between the sills of the two processes, $A_{\mathbf{x}t}$ and $e^{A_{\mathbf{x}t}}$, is the following:

$$\mathrm{var}(e^{A_{\mathbf{x}t}}) = \sigma_e^2 = \exp(\sigma_a^2\, 2) - \exp(\sigma_a^2), \tag{10.12}$$

where σ_e^2 is the sill for $e^{A_{\mathbf{x}t}}$.

Thus, by combining (10.11) and (10.12), we can easily obtain the value of the sill parameter for $Z_{\mathbf{x}t}$, the chlorophyll in the original scale. The same argument follows for the nugget parameter.

4.3   Effect of Sampling Resolution

In addition to position and direction, the semivariogram of a process $Z(\mathbf{x})$ can also depend on the spatial resolution of the sampling [Cla79]. Larger sample sizes tend to average out some of the smaller-scale variability in the data, thus reducing both the total variance and sill as well as changing the shape of the semivariogram at short distances. The modification of the semivariogram with the resolution of the measurements is termed *change of support* or *regularization*.

Consider the case where the coarse resolution estimate is simply an unweighted average of N higher-resolution samples:

$$\overline{Z(\mathbf{x})} = \frac{1}{N}\sum Z(\mathbf{x}). \tag{10.13}$$

If the high-resolution samples have equal variance σ_x^2, then the expected variance of the mean is

$$\sigma_{\bar{x}}^2 = \frac{1}{N}\sigma_x^2 + \frac{1}{N^2}\sum_i\sum_j\sigma_{ij}^2. \tag{10.14}$$

The variance of the mean ranges from $\sigma_{\bar{x}}^2 = \sigma_x^2$ for perfectly correlated samples $(\sigma_{ij}^2 = \sigma_x^2)$ to a minimum of σ_x^2/N for completely uncorrelated samples $(\sigma_{ij}^2 = 0)$. For intermediate regimes, the reduction of variance due to averaging can be estimated from the mean of the high-resolution semivariogram $\bar{\gamma}(\mathbf{v})$ over the averaging domain [Cla79]:

$$\sigma_{\bar{x}}^2 = \sigma_x^2 - \frac{1}{N^2}\left[\sum\sum\bar{\gamma}(\mathbf{v}_{ij})\right]. \tag{10.15}$$

Figure 6 shows a comparison of the north–south and east–west semivariograms computed from the Level 3 standard mapped image data ($\sim$9 km $\times$ 9 km) and the Bedford station Level 2 LAC data ($\sim$1 km $\times$ 1 km) for June for the box 30°W–35°W and 45°N–50°N. The total variance of the daily standard mapped image data is approximately 1/3 to 1/4 that of the original LAC data because of spatial averaging. The subsampling of the data from the fine resolution LAC grid to the medium resolution GAC grid ($\sim$4 km $\times$ 4 km) should not change the observed variance or semivariogram. Depending on the exact geometry of an individual satellite swath, the number of GAC points going into a single equal area, binned product grid cell, can vary from approximately 4 to 9. Given typical ranges of 30 km–50 km, the expected reduction in variance (excluding the nugget) is about 30%. In comparison, note that the variance would be expected to drop by a factor of about 6 for spatially uncorrelated data.

Temporal averaging on the binned product grid can contribute an additional reduction of variance in those regions, mostly subpolar, where more than one pass of a region is available for a single day. Finally, the variance will be reduced by the conversion of the binned product to the equal-angle, standard mapped image grid. The transformation involves a simple linear mapping between the two grids, but because of the distorted geometry of the binned product at high latitudes the remapping can lead to biases particularly in the zonal semivariogram estimates. The expected difference between the Level 2 and Level 3 standard mapped image data is difficult to quantify exactly because of the number of processing steps and the heterogeneity of the original sample coverage, and we are currently investigating how to calculate the semivariograms either from the original GAC data or from the daily binned product, which includes variance estimates. Other possible complications include overestimating the Level 2 variance from inclusion of outliers because of subtle differences in the data processing algorithms for Level 2 and Level 3 data.

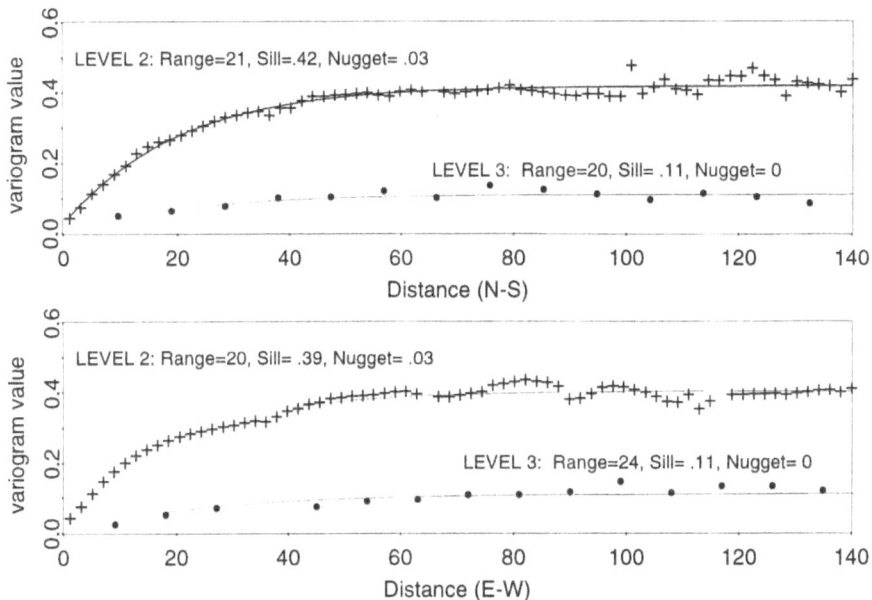

FIGURE 6. Comparison of the semivariograms for high resolution (Level 2) and low resolution (Level 3) satellite ocean chlorophyll for the region 30°W–35°W, 45°N–50°N for June 1998 in the north–south and east–west directions. Model parameters are estimated using nonlinear least squares.

5   Spatial Patterns for the North Atlantic Ocean

The variance and scales of the ocean color data are determined by the underlying physical and biological dynamics that govern phytoplankton distributions. One of the goals of our research is to see whether we can relate the spatial autocorrelation of ocean chlorophyll, defined by the range parameter of the fitted exponential semivariogram models, to patterns of physical variables. The analysis presented here is preliminary but points toward a complicated system with a number of competing effects.

The variation of the ocean color semivariogram with latitude from 25°N to 50°N is presented in Figure 7 for a set of 5° × 5° boxes along 35°W–30°W. Following the analysis above, each plot shows an exponential model fit to the median north–south semivariogram values for the 15 pairs of anomalies in June 1998. The range parameter decreases with latitude, going from about 69 km at 30°N–35°N to about 25 km at 50°N–55°N. The range is not well defined for the lowest latitude box 25°N–30°N, as shown by the observed nearly linear semivariogram and an extremely large standard error of 588 km on the model range fit. However, the standard error on the model range fit for the other boxes in Figure 7 is approximately 10 km. Figure 10 shows the range values for the semivariograms versus latitude along the meridional transect used in Figure 7. A similar, though slightly smaller, reduction in the estimated range is observed from the eastern to the western part of the North Atlantic Ocean as demonstrated in Figures 8 and 9, which cover $5° \times 5°$ boxes along 35°N–30°N from 20°W to 80°W. The model estimated range parameter decreases from approximately 50 km in the eastern subtropics to about 32 km in the west (Figure 10).

Significant work has been done on characterizing the mesoscale spatial variability of other physical variables. For example, [Smi99] using high-resolution, eddy-resolving numerical simulations of the North Atlantic, found that the zonal-average physical eddy autocorrelation spatial scales decreased from approximately 250 km at the Equator to about 40 km at 60°N. Similar spatial autocorrelation lengths have been calculated from satellite observations using TOPEX/Poseidon altimeter sea-surface height data poleward of 15°N [Fu,96] and from Advanced Very High Resolution Radiometer (AVHRR) sea-surface temperature imagery for the eastern North Atlantic from 35°N to 60°N [Kra90]. Krauss et al. [Kra90] and Stammer [Sta97] further show that the physical eddy spatial scales outside the tropics vary roughly linearly with the Rossby deformation radius; the horizontal scale where rotation becomes important relative to buoyancy and a key parameter governing turbulent ocean flow [Eme84]. These results are consistent with the hypothesis that the primary eddy-generation mechanism in the ocean is related to the baroclinic instability of the density field.

The ocean chlorophyll autocorrelation varies along the 30°W–35°W transect (Figure 10), showing a poleward decrease comparable to the one observed for the physical eddies ranges, suggesting that a similar eddy-generation mechanism may be responsible for the latitudinal variation in biological variables. The east to west decrease along 30°N–35°N (Figure 10) is more puzzling because the meridional variation of physical spatial scales and Rossby radius is much weaker than in the zonal direction [Eme84]. We speculate that the reduced spatial scales in the western basin may be related to western intensification of eddy kinetic energy associated with the Gulf Stream boundary current [Fu,96], [Sta97] and the subsequent rise in the importance of the sub-mesoscale (< 10 km) nutrient injection events and phytoplankton patchiness [Was98], [McG98].

6 Conclusions and Final Remarks

The availability of routine, high spatial resolution, satellite ocean color measurements provides a unique opportunity for characterizing the spatial and temporal scales of ocean biology. Complete knowledge of the space–time variability structure of the ocean color provides a window into the mechanisms of marine biological–physical interaction and are a necessary component of

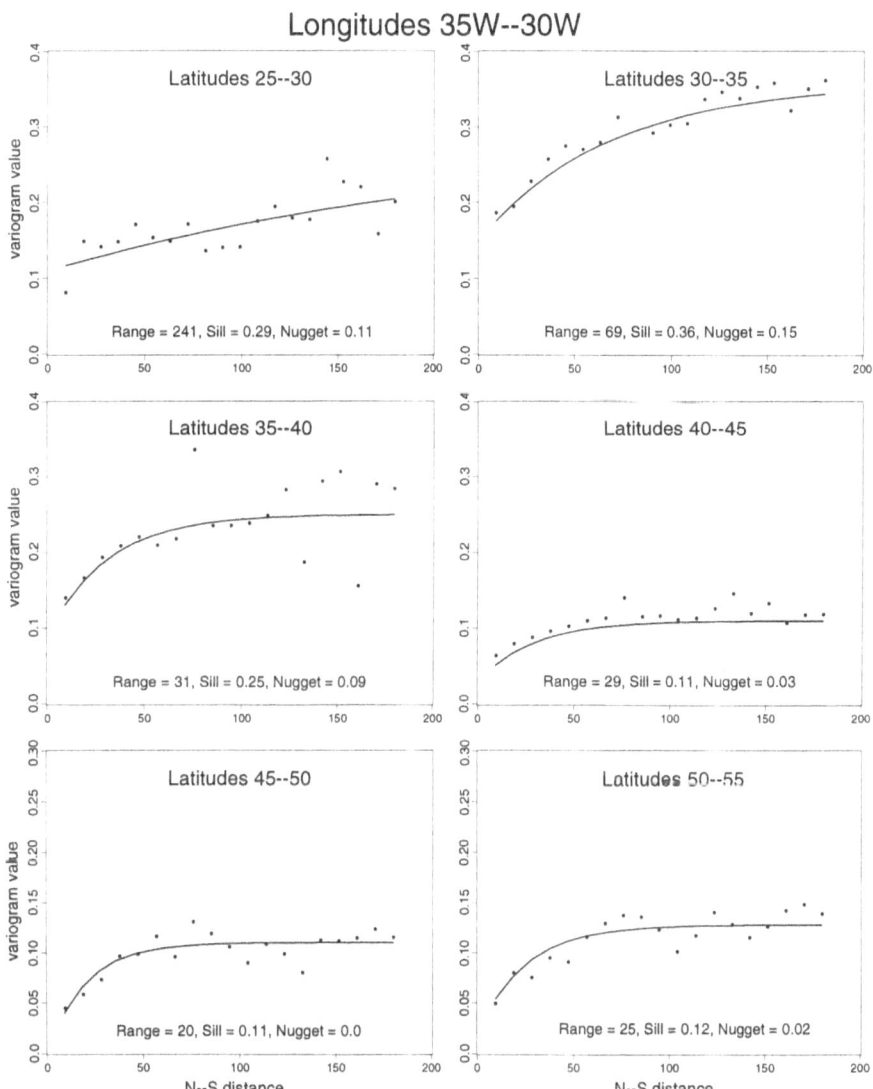

FIGURE 7. Semivariograms for the ocean chlorophyll anomalies in June 1998. Each panel shows the semivariogram value versus distance in the north–south direction for a 5° × 5° box along 35°W–30°W, with the latitude changing from panel to panel from 25°N to 55°N. Model parameters are estimated using nonlinear least squares.

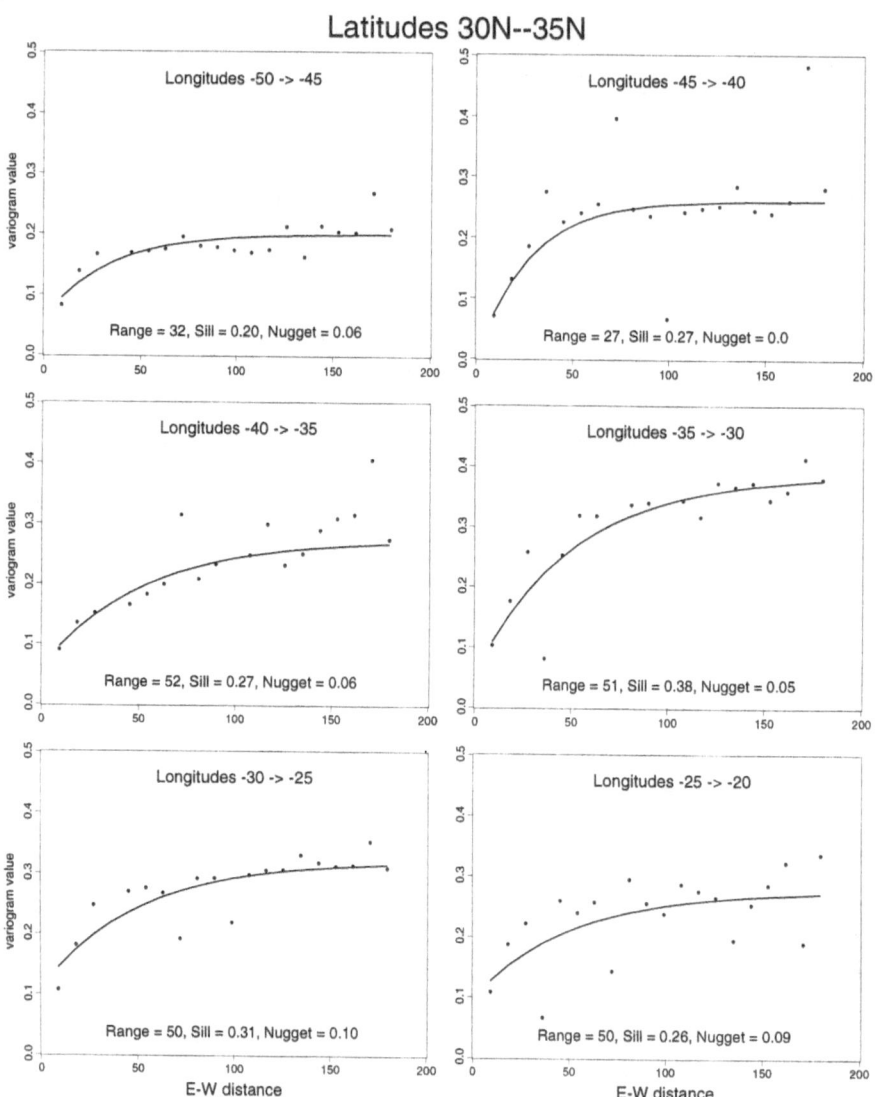

FIGURE 8. Semivariograms for the ocean chlorophyll anomalies in June 1998. Each panel shows the semivariogram value versus distance in the west–east direction for a 5° × 5° box along 30°N–35°N, with the longitude changing from panel to panel from 50°W to 20°W. Model parameters are estimated using nonlinear least squares.

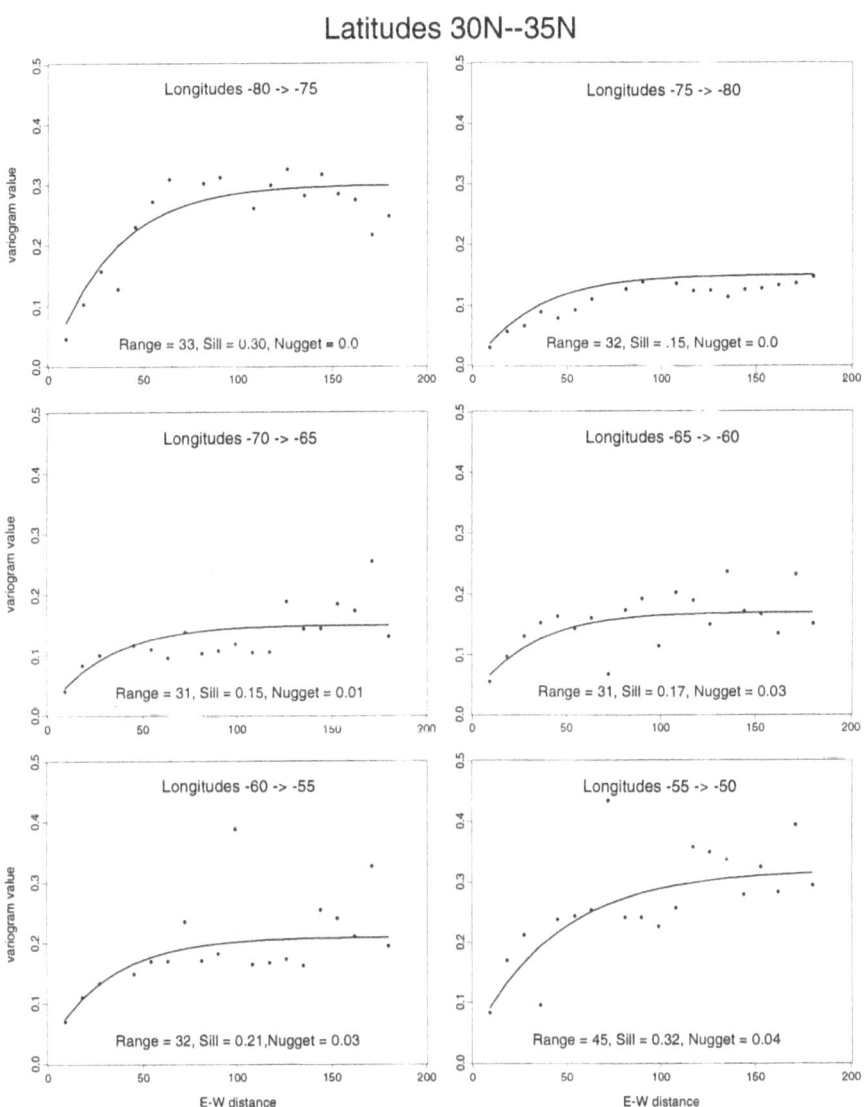

FIGURE 9. Semivariograms for the ocean chlorophyll anomalies in June 1998. Each panel shows the semivariogram value versus distance in the west–east direction for a 5° × 5° box along 30°N–35°N, with the longitude changing from panel to panel from 80°W to 50°W. Model parameters are estimated using nonlinear least squares.

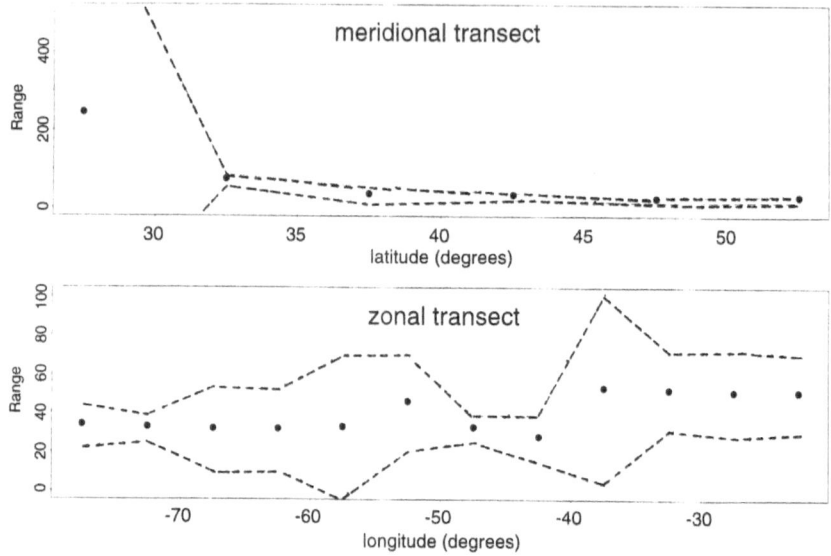

FIGURE 10. Range values for the semivariogram of the logarithmic transformation of the chlorophyll in a zonal and a meridional transect in the North Atlantic Ocean, in June 1998. The zonal transect is defined along 30°W–35°W and the meridional transect along 30°N–35°N. The dashed lines show confidence regions for the range parameter, using one standard deviation.

future work to objectively analyze satellite chlorophyll images using spatial predictors, such as kriging. We presented a detailed geostatistical framework for estimating spatial ranges of autocorrelation using the semivariogram and we showed preliminary calculations for a single month of the SeaWiFS data. The spatial patterns were found to be both nonstationary and anisotropic. In addition to the expected poleward decrease in spatial scales, as observed for physical variables, our analysis also suggested a significant east–west gradient that may be related to the elevated eddy kinetic energy in the Gulf Stream/North Atlantic Current system.

Further work is needed in a number of areas to clarify some of the questions raised here. Efforts are underway to expand our calculations to the full annual cycle and to the global domain to assess the robustness of the patterns we observe. We are also investigating the impact of the standard SeaWiFS data processing on variance and spatial autocorrelation scales. Finally, our treatment neglected any discussion of the temporal correlation scales. In addition to the obvious information gained on spatial and temporal scales of biological variability, a number of researchers (e.g., Chelton and Schlax [Che91]) propose local, temporal (rather than spatial) objective analysis as a simpler, more straightforward method of the interpolation of patchy satellite geophysical data such as ocean color.

Acknowledgments

We thank D. Nychka for his advice and comments and T. Hoar for his technical assistance. M. Fuentes acknowledges the support of the NCAR Geophysical Statistics Project, and S. Doney and D. Glover are supported in part by NASA SeaWiFS Grants W-19,223 and NAG-5-6456. This work would not be possible without the dedicated efforts of the NASA SeaWiFS project team (http://seawifs.gsfc.nasa.gov/SEAWIFS.html). Ocean color data used in this study were produced by the SeaWiFS Project at Goddard Space Flight Center. The data were obtained from the Goddard Distributed Active Archive Center. Use of this data is in accord with the SeaWiFS Research Data Use Terms and Conditions Agreement. We also thank the Bedford Institute of Oceanography, Nova Scotia, Canada, for supplying SeaWiFS data from their HRPT station. The National Center for Atmospheric Research is sponsored by the National Science Foundation.

References

[Abb98] Abbott, M.R. and Letelier, R.M. Decorrelation scales of chlorophyll as observed from bio-optical drifters in the California current. *Deep-Sea Research,* Part II, 45:1639–1667, 1998.

[Ahm87] Ahmed, S. and De Marsily, G. Comparison of geostatistical methods for estimating transmissivity using data on transmissivity and specific capacity. *Water Resources Research,* 23:1717–1737, 1987.

[Ali79] Ali, M.M. Analysis of stationary spatial-temporal processes: Estimation and prediction. *Biometrika,* 66:513–518, 1979.

[And98] Andrews, D.A. and Cox, D.D. Statistical modeling of vector-valued spatial data using gradient processes. David W. Scott, editor. In *Computing Science and Statistics 29: Proceedings of the 29th Symposium on the Interface,* pages 52–57. Interface Foundation of North America, Inc., Fairfax Station, VA, 1998.

[And99] Anderson, J.L. and Anderson, S.L. A Monte Carlo implementation of the non-linear filtering problem to produce ensemble assimilations and forecasts. *Monthly Weather Review,* submitted, 1999.

[Ben79] Bennett, R.J. *Spatial Time Series: Analysis-Forecasting-Control.* Pion, London, 1979.

[Ben98] Benjamin, S.G., Brown, J.M., Brundage K.J., Schwartz, B.E., and Smith, T.L. The operational RUC-2. *16th Conference on Weather Analysis and Forecasting,* Phoenix, AZ, 1998.

[Ber85] Berger, J.O. *Statistical Decision Theory and Bayesian Analysis.* Springer-Verlag, New York, 1985.

[Ber92] Berliner, L.M. Probability, statistics, and chaos. *Statistical Science,* 7:69–122, 1992.

[Ber96] Berliner, L.M. Hierarchical Bayesian time series models. In *Maximum Entropy and Bayesian Methods,* pages 15–22. Kluwer Academic, Dordrecht, 1996.

[Ber99a] Berliner, L.M. Monte Carlo based ensemble forecasting. *Statistics and Computing,* in press, 1999.

[Ber99b]   Berliner, L.M., Lu, Z.-Q., and Snyder, C. Statistical design for adaptive weather observations. *Journal of the Atmospheric Sciences*, 56:2536–2552, 1999.

[Ber99c]   Berliner, L.M., Royle, J.A., Wikle, C., and Milliff, R.F. Bayesian methods in the atmospheric sciences. J.M. Bernardo, J.O. Berger, A.P. Dawid, and A.F.M. Smith, editors. In *Bayesian Statistics 6*, pages 83–100. Oxford University Press, New York, 1999.

[Ber99d]   Berliner, L.M., Wikle, C.K., and Milliff, R.F. Multiresolution wavelet analyses in hierarchical Bayesian turbulence models. In *Bayesian Inference in Wavelet Based Models*. Lecture Notes in Statistics. Springer-Verlag, New York, 1999.

[Bil85]   Bilonick, R.A. The space–time distribution of sulfate deposition in the northeastern United States. *Atmospheric Environment*, 19:1829–1845, 1985.

[Bis99]   Bishop, C.H. and Toth, Z. Ensemble transformation and adaptive observations. *Journal of the Atmospheric Sciences*, 56:1748–1765, 1999.

[Box69]   Box, G.E.P. and Draper, N.R. *Evolutionary Operation: A Statistical Method for Process Improvement*. Wiley, New York, 1969.

[Box78]   Box, G.E.P., Hunter, W.G., and Hunter, J.S. *Statistics for Experimenters*. Wiley, New York, 1978.

[Bra99]   Brasseur, G.P., Orlando, J.J., and Tyndall, G.S., editors. *Atmospheric Chemistry and Global Change*. Oxford University Press, New York, 1999.

[Bro73]   Brown, R. New indices to locate clear-air turbulence. *Meteorological Magazine*, 102:347–2518, 1973.

[Bro97]   Brown, B.G. Status report on verification of turbulence algorithms. *Tech. Rep. RAP*, National Center for Atmospheric Research, 1997.

[Bui97]   Buizza, R. Potential forecast skill of ensemble prediction and spread and skill distributions of the ECMWF ensemble prediction system. *Monthly Weather Review*, 125:99–119, 1997.

[Bur98]   Burgers, G., Jan Van Leeuwen, P., and Evensen, G. Analysis scheme in the ensemble Kalman filter. *Monthly Weather Review*, 126:1719–1724, 1998.

[Cam95] Campbell, J.W. The lognormal distribution as a model for bio-optical variability in the sea. *Journal of Geophysical Research*, 100:13,237–13,254, 1995.

[Can96] Cane, M.A., Kaplan, A., Miller, R.N., Tang, B., Hackert, E.C., and Busalacchi, A.J. Mapping tropical Pacific sea level: Data assimilation via a reduced state space Kalman filter. *Journal of Geophsyical Research*, 101:22,599–22,617, 1996.

[Car96] Carlin, B.P. and Louis, T.A. *Bayes and Empirical Bayes Methods for Data Analysis.* Chapman and Hall, New York, 1996.

[Cas84] Caselton, W.F. and Zidek, J.V. Optimal monitoring network design. *Statistics and Probability Letters*, 2:223–227, 1984.

[Cas90] Casella, G. and Berger, R.L. *Statistical Inference.* Brooks/Cole (Wadsworth), Pacific Grove, CA, 1990.

[Cha95] Chaloner, K. and Verdinelli, I. Bayesian experimental design: A review. *Statistical Science*, 10:273–304, 1995.

[Che72] Chernoff, H. *Sequential Analysis and Optimal Design.* Society for Industrial and Applied Mathematics, Philadelphia, PA, 1972.

[Che91] Chelton, D.B. and Schlax, M.G. Estimation of time averages from irregularly spaced observations: With application to coastal zone color scanner estimates of chlorophyll concentration. *Journal of Geophysical Research*, 96:14,669–14,692, 1991.

[Che96a] Chen, S.S., Houze, R.A., and Mapes, B.E. Multiscale variability of deep convection in relation to large-scale circulation in TOGA COARE. *Journal of the Atmospheric Sciences*, 53(10):1380–1409, 1996.

[Che96b] Cherry, S. Singular value decomposition analysis and canonical correlation analysis. *J. Climate*, 9(9):2003–2009, 1996.

[Chi98] Chin, T.M., Milliff, R.F., and Large, W.G. Basin-scale, high-wavenumber sea surface wind fields from a multiresolution analysis of scatterometer data. *Journal of Atmospheric and Oceanic Technology*, 15:741–763, 1998.

[Chr91] Christensen, R. *Linear Models for Multivariate, Time Series, and Spatial Data.* Springer-Verlag, New York, 1991.

[Cla79] Clark, I. *Practical Geostatistics.* Elsevier Applied Science, New York, 1979.

[Cla93]   Clark, L.A. and Pregibon, D. Tree-based models. In *Statistical Models in S*, pages 377–419. Wadsworth & Brooks, Pacific Grove, CA, 1993.

[Cle79]   Cleveland, W.S. Robust locally weighted regression and smoothing scatterplots. *Journal of the American Statistical Association*, 74:829–836, 1979.

[Cli75]   Cliff, A.D. and Ord, J.K. Model building and the analysis of spatial pattern in human geography. *Journal of the Royal Statistical Society B*, 37:297–328, 1975.

[Coh93]   Cohen, A., Daubechies, I., and Vial, P. Wavelets on the interval and fast wavelet transforms. *Applied and Computational Harmonic Analysis*, 1:54–81, 1993.

[Coh97]   Cohn, S.E. An introduction to estimation theory. *Journal of the Meteorological Society of Japan*, 75:257–288, 1997.

[Col65]   Colson D. and Panofsky, H.A. An index of clear air turbulence. *Quarterly Journal of the Royal Meteorological Society*, 91:507–513, 1965.

[Cot89]   Cotton, W.R. and Anthes, R. *Storm and Cloud Dynamics*. International Geophysics Series, Vol. 44. Academic Press, New York, 1989.

[Cou97]   Courtier, P. Dual formulation of four-dimensional variational assimilation. *Quarterly Journal of the Royal Meteorological Society*, 123:2449–2461, 1997.

[Cre80]   Cressie, N.A.C. and Hawkins, D.M. Robust estimation of the variogram, I. *Journal of the International Association for Mathematical Geology*, 12:115–125, 1980.

[Cre85]   Cressie, N.A.C. When are relative variograms useful in geostatistics? *Journal of the International Association for Mathematical Geology*, 17:693–701, 1985.

[Cre90]   Cressie, N.A.C. The origins of kriging. *Mathematical Geology*, 22:239–252, 1990.

[Cre93]   Cressie, N.A.C. *Statistics for Spatial Data*. Wiley, New York, 1993.

[Cre96]   Cressie, N.A.C. Comment on "An approach to statistical spatial-temporal modeling of meteorological fields" by M.S. Handcock and J.R. Wallis. *Journal of the American Statistical Society*, 89:379–382, 1996.

[Cre98]   Cressie, N.A.C. and Wikle, C.K. Strategies for dynamic space–time statistical modeling: Discussion of "The Kriged Kalman filter" by K.V. Mardia et al. *Test*, 7:257–264, 1998.

[Cre99]   Cressie, N.A.C. and Huang, H. Classes of nonseparable, spatio-temporal stationary covariance functions. *Journal of the American Statistical Association*, 94:239–252, 1999.

[Cus94]   Cushman-Roisin, B. *Introduction to Geophysical Fluid Dynamics*. Prentice Hall, Englewood Cliffs, NJ, 1994.

[Dal91]   Daley, R. *Atmospheric Data Analysis*. Cambridge University Press, London, 1991.

[Dal97]   Daley, R. Atmospheric data assimilation. *Journal of the Meteorological Society of Japan*, 75:319–329, 1997.

[Dau92]   Daubechies, I. *Ten Lectures on Wavelets*. SIAM, Philadelphia, PA, 1992.

[Dav86]   Davis, J.C. *Statistics and Data Analysis in Geology*. Wiley, New York, 1986.

[Dig98]   Diggle, P.J., Moyeed, R.A., and Tawn, J.A. Model-based geostatistics (with discussion). *Applied Statistics*, 47:299–350, 1998.

[Don95]   Donoho, D.L., Johnstone, I.M., Kerkyacharian, G., and Picard, D. Wavelet shrinkage: Asymptotia. *Journal of the Royal Statistical Society, Series B*, 57:301–369, 1995.

[Dut80]   Dutton, M.J.O. Probability forecasts of clear-air turbulence based on numerical output. *Meteorological Magazine*, 109:293–310, 1980.

[Ell85]   Ellrod, G.P. Detection of high level turbulence using satellite imagery and upper air data. *NOAA Tech. Memo NESDIS*, 10:1–30, 1985.

[Ell92a]   Ellner, S., Nychka, D.W., and Gallant, A.R. LENNS, a program to estimate the dominant Lyapunov exponent of noisy nonlinear systems from time series data. Institute of Statistics Mimeo Series 2235, Statistics Department, North Carolina State University, Raleigh, NC 27695-8203, 1992.

[Ell92b]   Ellrod, G.P. and Knapp, D.I. An objective clear-air turbulence forecasting technique: Verification and operational use. *Weather and Forecasting*, 7:150–165, 1992.

[Ema94]   Emanuel, K.A. *Atmospheric Convection*. Oxford University Press, New York, 1994.

[Eme84]   Emery, W.J., Lee, W.G., and Magaard, L. Geographic and seasonal distributions of Brunt-Vaisala frequency and Rossby radii in the North Pacific and North Atlantic. *Journal of Physical Oceanography*, 14:294–317, 1984.

[End64]   Endlich, R.M. The mesoscale structure of some regions of clear air turbulence. *Journal of Applied Meteorology*, 3:261–276, 1964.

[Eps85]   Epstein, E.S. *Statistical Inference and Prediction in Climatolgy: A Bayesian Approach*. American Meteorological Society, Boston, MA, 1985.

[Err99]   Errico, R.M., Fillion, L., Nychka, D., and Lu, Z.-Q. Some statistical considerations associated with the data assimilation of precipitation observations. *Quarterly Journal of the Royal Meteorological Society*, in press, 1999.

[Fed72]   Fedorov, V.V. *Theory of Optimal Experiments*. Academic Press, New York, 1972.

[Fed97]   Fedorov, V. Placement of sensors for correlated observations. E.J. Wegman and S.P. Azen, editors. In *Computing Science and Statistics 29: Computational Statistics and Data Analysis*, pages 110–114. Interface Foundation of North America, Inc., Fairfax Station, VA, 1997.

[Fis95]   Fisher, M. and Courtier, P. Estimating the covariance matrices of analysis and forecast error in variational data assimilation. Technical memorandum 220, ECMWF Technical Memorandum, 1995.

[Fri91]   Friedman, J.H. Multivariate adaptive regression splines. *Annals of Statistics*, 19:1–141, 1991.

[Fu,96]   Fu, L.-L. and Smith, R.D. Global ocean circulation from satellite altimetry and high-resolution computer simulation. *Bulletin of the American Meteorological Society*, 77:2625–2636, 1996.

[Gan65]   Gandin, L. *Objective analysis of meteorological fields* (1963, in Russian). English translation by Israel Program for Scientific Translations, Jerusalem, 1965.

[Ghi81]   Ghil, M., Cohn, S.E., Tavantzis, J., Bube, K., and Isaacson, E. Applications of estimation theory to numerical weather predictions, In *Dynamic meteorology: Data assimilation methods*, pages 139–224. Springer-Verlag, New York, 1981.

[Gil96]   Gilks, W.R., Richardson, S., and Spiegelhalter, D.J., editors. *Markov Chain Monte Carlo in Practice*. Chapman and Hall, London, 1996.

[Goo94]   Goodall, C. and Mardia, K.V. Challenges in multivariate spatio-temporal modeling. In *Proceedings of the XVIIth International Biometric Conference*, Hamilton, Ontario, Canada, 1994.

[Got96]   Gotway, C.A. and Hartford, A.H. Geostatistical methods for incorporating auxiliary information in the prediction of spatial variables. *Journal of Agricultural, Biological and Environmental Statistics*, 1:17–39, 1996.

[Gre94]   Green, P.J. and Silverman, B.W. *Nonparametric Regression and Generalized Linear Models, A Roughness Penalty Approach*. Chapman and Hall, London, 1994.

[Gut94]   Guttorp, P., Meiring, W., and Sampson, P.D. A space–time analysis of ground-level ozone data. *Environmetrics*, 5:241–254, 1994.

[Haa90]   Haas, T.C. Lognormal and moving-window methods of estimating acid deposition. *Journal of the American Statistical Association*, 85:950–963, 1990.

[Haa95]   Haas, T.C. Local prediction of a spatio-temporal process with an application to wet sulfate deposition. *Journal of the American Statistical Association*, 90:1189–1199, 1995.

[Haa96]   Haas, T.C. Multivariate spatial prediction in the presence of nonlinear trend and covariance non-stationarity. *Environmetrics*, 7:145–165, 1996.

[Ham99]   Hamill, T.M., Snyder, C., and Morss, R.E. A comparison of probabilistic forecasts from bred, singular vector and perturbed observation ensembles. *Monthly Weather Review*, submitted, 1999.

[Har92]   Harvey, L.O., Hammond, K.R., Lusk, C.M., and Mross, E.F. The application of signal detection theory to weather forecasting behavior. *Monthly Weather Review*, 120:863–883, 1992.

[Has94]   Hastie, T., Tibshirani, R., and Buja, A. Flexible discriminant analysis by optimal scoring. *Journal of the American Statistical Association*, 89:1255–1270, 1994.

[Hau78]   Haury, L.R., McGowan, J.A., and Wiebe, P.H. Patterns and processes in the time–space scales of plankton distributions. In *Spatial Patterns in Plankton Communities*, pages 277–327. Plenum Press, New York, 1978.

[Hof94]   Hofmann, D.J., Oltmans, S.J., Komhyr, W.D., Harris, J.M., Lath-rop, J.A., Langford, A.O., Deshler T., Johnson B.J., Torres, A., and Matthews, W.A. Ozone loss in the lower stratosphere over the United States in 1992–1993: Evidence for heterogeneous chemistry on the pinatubo aerosol. *Geophysical Research Letters*, 21:65–68, 1994.

[Hol92]   Holton, J.R. *An Introduction to Dynamic Meteorology*, 3rd ed. Academic Press, San Diego, 1992.

[Hol95]   Hollandworth, S.M., Bowman, K.P., and McPeters, R.D. Observational study of the quasi-biennial oscillation in ozone. *Journal of Geophysical Research. Atmospheres*, 100:7347–7361, 1995.

[Hoo93]   Hooker, S.B. and Esaias, W.E. An overview of the SeaWiFS project. *EOS, Transactions of the American Geophysical Union*, 74:241–246, 1993.

[Hou98]   Houtekamer, P.L. and Mitchell, H.L. Data assimilation using an ensemble Kalman filter technique. *Monthly Weather Review*, 126:796–811, 1998.

[Hua96]   Huang, H.C. and Cressie, N.A.C. Spatio-temporal prediction of snow water equivalent using the Kalman filter. *Computational Statistics and Data Analysis*, 22:159–175, 1996.

[Jaz70]   Jazwinski, A.H. *Stochastic Processes and Filtering Theory*. Academic Press, New York, 1970.

[Jol86]   Jolliffe, I.T. *Principal Component Analysis*. Springer Series in Statistics. Springer-Verlag, New York, 1986.

[Jol97]   Joly, A., Jorgensen, D., Shapiro, M.A., Thorpe, A., Bessemoulin, P., Browning, K.A., Cammas, J.-P., Chalon, J.-P., Clough, S.A., Emanuel, K.A., Eymard, L., Gall, R., Hildebrand, P.H., Langland, R.H., Lemaître, Y., Lynch, P., Moore, J.A., Persson, P., Snyder, C., and Wakimoto, R.M. Definition of the fronts and Atlantic storm-track experiment (FASTEX). *Bulletin of American Meteorological Society*, 78:1917–1940, 1997.

[Jou78]   Journel, A. and Huijbregts, C. *Mining Geostatistics*. Academic Press, New York, 1978.

[Kag97]   Kagan, R.L. *Averaging of Meteorological Fields* (1979, in Russian). Kluwer Academic, Dordrecht, 1997.

[Kal60a] Kalman, R.E. A new approach to linear filtering and prediction problems. *Transactions of the ASME Journal of Basic Engineering*, 82:35–45, 1960.

[Kal60b] Kalman, R.E. and Bucy, R.S. New results in linear filtering and prediction theory. *Transactions of the ASME Journal of Basic Engineering*, 83:95–108, 1960.

[Kol41a] Kolmogorov, A.N. The local structure of turbulence in incompressible viscous fluid for very large Reynolds numbers. *Doklady Akademii Nauk SSSR*, 30:301–305, 1941.

[Kol41b] Kolmogorov, A.N. On degeneration of isotropic turbulence in an incompressible viscous liquid. *Doklady Akademii Nauk SSSR*, 31:538–541, 1941.

[Kra90] Krauss, W., Doscher, R., Lehmann, A. and Viehoff, T. On eddy scales in the eastern and northern North Atlantic Ocean as a function of latitude. *Journal of Geophysical Research*, 95:18,049–18,056, 1990.

[Kri51] Krige, D.G. A statistical approach to some basic mine valuation problems on the Witwaterstrand. *Journal of the Chemical, Metallurgical and Mining Society of South Africa*, 52:119–139, 1951.

[Kyr99] Kyriakidis, P.C. and Journel, A.G. Geostatistical space–time models: A review. *Mathematical Geology*, 31:651–684, 1999.

[Le,94] Le, N.D. and Zidek, J.V. Network design for monitoring multivariate random spatial fields. J.P. Vilaplana and M.L. Puri, editors. In *Recent Advances in Statistics and Probability*, pages 191–206. VSP, Zeist, 1994.

[Lei74] Leith, C.E. Theoretical skill of Monte Carlo forecasts. *Monthly Weather Review*, 102:409–418, 1974.

[Lev99] Levine, R.A. and Berliner, L.M. Statistical principles for climate change studies. *Journal of Climate*, 12:564–574, 1999.

[Log94] Logan, J.A. Trends in the vertical distribution of ozone: An analysis of ozonesonde data. *Journal of Geophysical Research*, 99:25,553–25,585, 1994.

[Lor86] Lorenc, A.C. Analysis methods for numerical weather prediction. *Quarterly Journal of the Royal Meteorological Society*, 112:1177–1194, 1986.

[Lor93]   Lorenz, E.N. *The Essence of Chaos.* University of Washington Press, Seattle, WA, 1993.

[Lor98]   Lorenz, E.N. and Emanuel, K.A. Optimal sites for supplementary weather observations: Simulation with a small model. *Journal of the Atmospheric Sciences*, 55:399–414, 1998.

[Lu,99]   Lu, Z.-Q. and Berliner, L.M. Markov switching time series models with application to a daily runoff series. *Water Resources Research*, 35:523–534, 1999.

[Maj97]   Majure, J.J. and Cressie, N. Dynamic graphics for exploring spatial dependence in multivariate spatial data. *Geographical Systems*, 4:131–158, 1997.

[Mal89]   Mallat, S.G. A theory for multiresolution signal decomposition: The wavelet representation. *IEEE Transactions on Pattern Analysis and Machine Intelligence*, 11:674–693, 1989.

[Mar75]   Martin, R.L. and Oeppen, J.E. The identification of regional forecasting models using space–time correlation functions. *Transactions of the Institute of British Geographers*, 66:95–118, 1975.

[Mar79]   Mardia, K.V., Kent, J.T., and Bibby, J.M. *Multivariate Analysis.* Academic Press, London, 1979.

[Mar94]   Marroquin, A. An advanced algorithm to diagnose atmospheric turbulence using numerical model output. *Geophysical Research Letters*, 21:2515–2518, 1994.

[Mar98]   Mardia, K.V., Goodall, C., Redfern, E.J., and Alonso, F.J. The kriged Kalman filter. *Test*, 7:217–285, 1998.

[Mat71]   Matheron, G. The theory of regionalized variables and its applications. Fontainebleau, Les Cahiers du Centre de Morphologie Mathématique, 1971.

[Mat86]   Matern, B. *Spatial Variation.* Lectures Notes in Statistics, Vol. 36. Springer-Verlag, New York, 1986.

[McC97]   McCann, D.W. MWAVE—An algorithm to diagnose breaking mountain waves. *Weather and Forecasting* (submitted), 1997.

[McC98]   McClain, C.R., Cleave, M.L., Feldman, G.C., Gregg, W.W., Hooker, S.B., and Kuring, N. Science quality SeaWiFS data for global biosphere research. *Sea Technology*, 39:10–14, 1998.

[McG98] McGillicuddy, D.J. Jr., Robinson, A.R., Siegel, D.A., Jannasch, H.W., Johnson, R., Dickey, T.D., McNeil, J., Michaels, A.F., and Knap, A.H. Influence of mesoscale eddies on new production in the Sargasso Sea. *Nature*, 394:263–266, 1998.

[McI91] McIlveen, J.F.R. *Fundamentals of Weather and Climate*. Chapman and Hall, London, 1991.

[Mei98a] Meiring, W. and Nychka, D.W. Functional data analysis for vertical profiles. In S. Weisberg, editor, *Interface Proceedings: Computing Science and Statistics*, volume 30, pages 115–133. Interface Foundation of North America, Inc., Fairfax Station, VA, 1998.

[Mei98b] Meiring, W., Guttorp, P., and Sampson, P.D. Space–time estimation of grid-cell hourly ozone levels for assessment of a deterministic model. *Environmental and Ecological Statistics*, 5:197–222, 1998.

[Mej74] Mejia, J.M. and Rodriguez-Iturbe, I. Correlation links between normal and log normal processes. *Water Resources Research*, 10:689–690, 1974.

[Mey93] Meyer, Y. *Wavelets: Algorithms and Applications*. SIAM, Philadelphia, PA, 1993.

[Mil96] Milliff, R.A., Large, W.G., Holland, W.R., and McWilliams, J.C. The general circulation responses of high-resolution North Atlantic Ocean models to synthetic scatterometer winds. *Journal of Physical Oceanography*, 26:1747–1768, 1996.

[Mil97a] Miller A.J., Flynn, L.E., Hollandsworth, S.M., DeLuisi, J.J., Petropavlovskikh, I.V., Tiao, G.C., Reinsel, G.C., Wuebbles, D.J., Kerr, J., Nagatani, R.M., Bishop, L., and Jackman, C.H. Information content of umkehr and solar backscattered ultraviolet (SBUV) 2 satellite data for ozone trends and solar responses in the stratosphere. *Journal of Geophysical Research*, 102:19,257–19,263, 1997.

[Mil97b] Miller, R.N., Busalacchi, A.J., and Hackert, E.C. Applications of data assimilation to analysis of the ocean on large scales. *Journal of the Meteorological Society of Japan*, 75:445–462, 1997.

[Mon95] Moncrieff, M.W. Mesoscale convection from a large-scale perspective. *Atmospheric Research*, 35:87–112, 1995.

[Moo92] Moorthi, S. and Suarez, M.J. Relaxed Arakawa–Schubert: A parameterization of moist convection for general circulation models. *Quarterly Journal of the Royal Meteorological Society*, 120:978–1002, 1992.

[Mye82] Myers, D.E. Matrix formulation of co-kriging. *Mathematical Geology*, 14:249–257, 1982.

[Mye91] Myers, D.E. Pseudo cross-variograms, positive-definiteness, and cokriging. *Mathematical Geology*, 23:805–816, 1991.

[Nas93] Nason, G.P. The **wavethresh** package: Wavelet transfrom and thresholding software for S. Available from the StatLib archive, 1993.

[Nyc] Nychka, D.W., Bailey, B.A., Ellner, S., Haaland, P., and O'Connell, M. *FUNFITS: Data analysis and statistical tools for estimating functions*. Department of Statistics, North Carolina State University, Raleigh, NC 27695-8203.

[Nyc96] Nychka, D., Yang, Q., and Royle, J.A. Constructing spatial designs using regression subset selection. V. Barnett and K.F. Turkman, editors, In *Statistics for the Environment 3: Sampling and the Environment*, pages 131–154. Wiley, New York, 1996.

[Nyc99] Nychka, D.W., Wikle, C.K., and Royle, J.A. Large spatial prediction problems and nonstationary random fields. *Journal of the Royal Statistical Society B*, submitted, 1999.

[Oeh93] Oehlert, G.W. Regional trends in sulfate wet deposition. *Journal of the American Statistical Association*, 88:390–399, 1993.

[O'R98] O'Reilly, J.E., Maritorena, S., Mitchell, B.G., Siegel, D.A., Carder, K.L., Garver, S.A., Kahru, M., and McClain, C. Ocean color chlorophyll algorithms for SeaWiFS. *Journal of Geophysical Research*, 103:24,937–24,953, 1998.

[Pal98] Palmer, T.N., Gelaro, R., Barkmeijer, V., and Buizza, R. Singular vectors, metrics, and adaptive observations. *Journal of the Atmospheric Sciences*, 55:633–653, 1998.

[Pap65] Papoulis, A. *Probability, Random Variables, and Stochastic Processes*. McGraw-Hill, New York, 1965.

[Pei92] Peixoto, J.P. and Oort, A.H. *Physics of Climate*. American Institute of Physics, New York, 1992.

[Pen93] Penland, C. and Magorian, T. Prediction of el niño sea surface temperatures using linear inverse modeling. *Journal of Climate*, 6:1067–1076, 1993.

[Pfe80] Pfeifer, P.E. and Deutsch, S.J. Identification and interpretation of first-order space–time ARMA models. *Technometrics*, 22:397–408, 1980.

[Pre88] Preisendorfer, R.W. *Principal Component Analysis in Meteorology and Oceanography*. Developments in Atmospheric Science. Elsevier, New York, 1988.

[Puk93] Pukelsheim, F. *Optimal Design of Experiments*. Wiley, New York, 1993.

[Ram97] Ramsay, J.O. and Silverman, B.W. *Functional Data Analysis*. Springer-Verlag, New York, 1997.

[Ran96] Randel, W.J. and Wu, F. Isolation of the ozone QBO in SAGE II data by singular-value decomposition. *Journal of the Atmospheric Sciences*, 53:2546–2559, 1996.

[Rea96] Reap, R.M. Probability forecasts of clear-air turbulence for the contiguous U.S. *Bulletin of the National Weather Service Office of Meteorology Technical Procedures*, 430:1–15, 1996.

[Rod74] Rodriguez-Iturbe, I. The design of rainfall networks in space and time. *Water Resources Research*, 10:713–728, 1974.

[Rou89] Rouhani, S. and Hall, T.J. Space–time kriging of groundwater data. M. Armstrong, editor, *Geostatistics*, Vol. 2, pages 639–650. Kluwer Academic, Dordrecht, 1989.

[Rou90a] Rouhani, S. and Myers, D.E. Problems in space–time kriging of geohydrological data. *Mathematical Geology*, 22:611–623, 1990.

[Rou90b] Rouhani, S. and Wackernagel, H. Multivariate geostatistical approach to space–time data analysis. *Water Resources Research*, 26:585–591, 1990.

[Roy99a] Royle, J.A. Multivariate spatial models. L.M. Berliner, D. Nychka and T. Hoar, editors. In *In Studies in the Atmospheric Sciences*, pages 23–44. Springer-Verlag, New York, 1999.

[Roy99b] Royle, J.A. and Berliner, L.M. A hierarchical approach to multivariate spatial modeling and prediction. *Journal of Agricultural, Biological, and Environmental Statistics*, 4:29–56, 1999.

[Roy99c] Royle, J.A., Berliner, L.M., Wikle, C.K., and Milliff, R. A hierarchical spatial model for constructing wind fields from scatterometer data in the Labrador Sea. C. Gatsonis, R.E. Kass, B. Carlin, A. Carriquiry, A. Gelman, I. Verdinelli, M. West, editors. In *Case Studies in Bayesian Statistics,* Volume IV, pages 367–382. Springer-Verlag, New York, 1999.

[Sac89] Sacks, J., Welch, W.J., Mitchell, T.J., and Wynn, H.P. Design and analysis of computer experiments. *Statistical Science,* 4:409–435, 1989.

[Sci93] Scipione, C.M. and Berliner, L.M. Bayesian statistical inference in nonlinear dynamical systems. In *Proceedings of the Bayesian Section of the American Statistical Association,* pages 73–78, Alexandria, VA, 1993.

[Sha98] Sharman, R. and Cornman L. An integrated approach to clear-air turbulence prediction. *36th Aerospace Sciences Meeting & Exhibit,* Reno, NV, 1998.

[Shu82] Shumway, R.H. and Stoffer, D.J. An approach to time series smoothing and forecasting using the em algorithm. *Journal of Time Series Analysis,* 3:253–264, 1982.

[Shu88] Shumway, R.H. *Applied Statistical Time Series Analysis.* Prentice Hall, Englewood Cliffs, NJ, 1988.

[Sil80] Silvey, S.D. *Optimal Design.* Chapman and Hall, New York, 1980.

[Sil96] Silverman, B.W. Smoothed functional principal components analysis by choice of norm. *Annals of Statistics,* 24:1–24, 1996.

[Sim96] Simonoff, J.S. *Smoothing Methods in Statistics.* Springer-Verlag, New York, 1996.

[Smi99] Smith, R.D., Maltrud, M.E., Bryan, F.O., and Hecht, M.W. Numerical simulation of the North Atlantic at $1/10°$. *Journal of Physical Oceanography,* submitted, 1999.

[Sol96] Solomon, S., Portmann, R.W., Garcia R.R., Thomason, L.W., Poole, L.R., and McCormick, M.P. The role of aerosol variations in anthropogenic ozone depletion at northern mid-latitudes. *Journal of Geophysical Research,* 101:22,991–23,006, 1996.

[Sol98] Solomon, S., Portmann, R.W., Garcia, R.R., Randel, W., Wu, F., Nagatani, R., Gleason, J., Thomason, L., Poole, L.R., and McCormick, M.P. Ozone depletion at mid-latitudes: Coupling of volcanic aerosols and temperature variability to anthropogenic chlorine. *Geophysical Research Letters,* 25:1871–1874, 1998.

[SRN98]   World Meteorological Organization Ozone Research SPARC Report
          No. 1 and Monitoring Project Report Number 43. *Assessment of
          Trends in the Vertical Distribution of Ozone.* The World Climate
          Research Programme Project on Stratospheric Processes and Their
          Role in Climate (SPARC), the International Ozone Commission
          (IOC), and the World Meteorological Organization Global Atmo-
          spheric Watch programme (GAW), 1998.

[Sta97]   Stammer, D. Global characteristics of ocean variability estimated
          from regional topex/poseidon altimeter measurements. *Journal
          Physical Oceanography*, 27:1743–1769, 1997.

[Ste78]   Steele, J.H. *Spatial Patterns in Plankton Communities.* Plenum
          Press, New York, 1978.

[Swi79]   Switzer, P. Statistical considerations in network design. *Water Re-
          sources Research*, 15:1712–1716, 1979.

[Tal97]   Talagrand, O. Assimilation of observations, An introduction. *Jour-
          nal of the Meteorological Society of Japan*, 75:191–209, 1997.

[Tar87]   Tarantola, A. *Inverse Problem Theory: Methods for Data Fitting
          and Model Parameter Estimation.* Elsevier, New York, 1987.

[Thi85]   Thiebaux, H.J. On approximations to geopotential and wind-field
          correlation structures. *Tellus*, 37A:126–131, 1985.

[Tia90]   Tiao, G.C., Reinsel, G.C., Xu, D., Pedrick, J.H., Zhu, X., Miller,
          A.J., DeLuisi, J., Mateer, C.L., and Wuebbles, D.J. Effects of au-
          tocorrelation and temporal sampling schemes on estimates of trend
          and spatial correlation. *Journal of Geophysical Research*, 95:20,507–
          20,517, 1990.

[Tit80]   Titterington, D.M. Aspects of optimal design in dynamic systems.
          Technometrics, 22:287–299, 1980.

[Tot97]   Toth, Z. and Kalnay, E. Ensemble forecasting at NCEP and the
          breeding method. *Monthly Weather Review*, 125:3297–3318, 1997.

[Tre92]   Trenberth, K.E. (editor). *Climate System Modeling.* Cambridge
          University Press, London, 1992.

[Vel94]   Velden, C.S. and Young, J.A. Satellite observations during TOGA
          COARE: Large-scale descriptive overview. *Monthly Weather Re-
          view*, 122:2426–2441, 1994.

[Ver93]   Ver Hoef, J.M. and Cressie, N.A.C. Multivariable spatial prediction.
          Mathematical Geology, 25:219–239, 1993.

[Ver98]   Ver Hoef, J.M. and Barry, R.P. Modeling crossvariograms for co-kriging and multivariable spatial prediction. *Journal of Statistical Planning and Inference*, 69:275–294, 1998.

[von95]   von Storch, H., Burger, G., Schnur, R., and von Storch, J.S. Principal oscillation patterns: A review. *Journal of Climate*, 8:377–400, 1995.

[von98]   von Storch, H. and Zwiers, F.W. *Statistical Analysis in Climate Research*. Cambridge University Press, Cambridge, 1998.

[Wac95]   Wackernagel, H. *Multivariate Geostatistics*. Springer-Verlag, Berlin, 1995.

[Wah90]   Wahba, G. *Spline Models for Observational Data*. SIAM, Philadelphia, PA, 1990.

[Was98]   Washburn, L., Emery, B.M., Jones, B.H., and Ondercin, D.G. Eddy stirring and phytoplankton patchiness in the subartic North Atlantic in late summer. *Deep-Sea Research I*, 45:1411–1439, 1998.

[Wea98]   Weatherhead, E.C., Reinsel, G.C., Tiao, G.C., Meng, X-L., Choi, D., Cheang, W-K., Keller, T., DeLuisi, J., Wuebbles, D.J., Kerr, J.B., Miller, A.J., Oltmans, S.J., and Frederick, J.E. Factors affecting the detection of trends: Statistical considerations and applications to environmental data. *Journal of Geophysical Research*, 103:17,149–17,161, 1998.

[Wes97]   West, M. and Harrison, J. *Bayesian Forecasting and Dynamic Models*, 2nd ed. Springer-Verlag, New York, 1997.

[Whi54]   Whittle, P. On stationary processes in the plane. *Biometrika*, 41:434–449, 1954.

[Wik96]   Wikle, C.K. *Spatio-temporal statistical models with applications to atmospheric processes*. PhD thesis, Iowa State University, Ames, IO, 1996.

[Wik97]   Wikle, C.K. and Cressie, N.A.C. A dimension reduction approach to space–time Kalman filtering. Preprint 97-24, Statistical Laboratory, Iowa State University, Ames, IO, 1997.

[Wik98a]   Wikle, C.K., Berliner, L.M., and Cressie, N.A.C. Hierarchical Bayesian space–time models. *Environmental and Ecological Statistics*, 5:117–154, 1998.

[Wik98b] Wikle, C.K., Milliff, R.F., Nychka, D., and Berliner, L.M. Spatio-temporal hierarchical Bayesian blending of tropical ocean surface wind data. Technical Report GSP98-01, Geophysical Statistics Project, National Center for Atmospheric Research, Boulder, CO, 1998.

[Wik99a] Wikle, C.K. Hierarchical space–time dynamic models. L.M. Berliner, D. Nychka and T. Hoar, editors. In *Studies in the Atmospheric Sciences*, pages 45–64. Springer-Verlag, New York, 1999.

[Wik99b] Wikle, C.K. and Cressie, N.A.C. Space–time statistical modeling of environmental data. In *Quantifying Spatial Uncertainty in Natural Resources: Theory and Applications for GIS and Remote Sensing*, pages 213–236. Ann Arbor Press, Chelsea, MI, 1999.

[Wik99c] Wikle, C.K. and Royle, J.A. Space–time dynamic design of environmental monitoring networks. *Journal of Agricultural, Biological, and Environmental Statistics*, 1999. To appear.

[Wik99d] Wikle, C.K., Milliff, R.F., and Large, W.G. Surface wind variability on spatial scales from 1 to 1000 km observed during TOGA COARE. *Journal of the Atmospheric Sciences*, 56:2222–2231, 1999.

[Wik99e] Wikle, C.K., Milliff, R.F., Nychka, D., and Berliner, L.M. Spatio-temporal hierarchical Bayesian modeling: Tropical ocean surface winds. *Journal of the American Statistical Society*, 1999. Pending review.

[Wil95] Wilks, D.S. *Statistical Methods in the Atmospheric Sciences*. Academic Press, San Diego, 1995.

[Wol78] Wold, S. Cross-validatory estimation of the number of components in factor and principal components models. *Technometrics*, 20:397–405, 1978.

[Wor93] Wornell, G.W. Wavelet-based representations for the 1/F family of fractal processes. *Proceedings of the IEEE*, 81:1428–1450, 1993.

[Wor95] World Meteorological Organization Global Ozone Research and Monitoring Project—Report No. 37. *Scientific Assessment of Ozone Depletion: 1994*. National Oceanic and Atmospheric Administration, National Aeronautics and Space Administration, United Nations Environment Programme, World Meteorological Organization, 1995.

[Wro89] Wroblewski, J. A model of the spring bloom in the North Atlantic and its impact on ocean optics. *Limnology and Oceanography*, 34:1563–1571, 1989.

[Wu,96]   Wu, X. and Moncrieff, M. New insights and approaches to convective parameterization. In *Recent Progress on Cloud-Resolving Modeling of TOGA COARE and GATE Cloud Systems*, UK, 1996. ECMWF.

[Yan99]   Yano, J., Moncrieff, M., Wu, X., and Yamada, M. Wavelet decomposition of the spatial structure associated with mesoscale organized scale, submitted, 1999.

[Yod87]   Yoder, J.A., McClain, C.R., Blanton, J.O., and Oey, L.-Y. Spatial scales in czcs-chlorophyll imagery of the southeastern U.S. continental shelf. *Limnology and Oceanography*, 32:929–941, 1987.

[Zim92]   Zimmerman, D.L. and Cressie, N.A.C. Mean squared prediction error in the spatial linear model with estimated covariance parameters. *Annals of the Institute of Statistical Mathematics*, 44:27–43, 1992.

[Zim93]   Zimmerman, D.L. Another look at anisotropy in geostatistics. *Mathematical Geology*, 25:453–470, 1993.

Index

Lecture Notes in Statistics

For information about Volumes 1 to 70,
please contact Springer-Verlag

Vol. 71: E.M.R.A. Engel, A Road to Randomness in Physical Systems. ix, 155 pages, 1992.

Vol. 72: J.K. Lindsey, The Analysis of Stochastic Processes using GLIM. vi, 294 pages, 1992.

Vol. 73: B.C. Arnold, E. Castillo, J.-M. Sarabia, Conditionally Specified Distributions. xiii, 151 pages, 1992.

Vol. 74: P. Barone, A. Frigessi, M. Piccioni, Stochastic Models, Statistical Methods, and Algorithms in Image Analysis. vi, 258 pages, 1992.

Vol. 75: P.K. Goel, N.S. Iyengar (Eds.), Bayesian Analysis in Statistics and Econometrics. xi, 410 pages, 1992.

Vol. 76: L. Bondesson, Generalized Gamma Convolutions and Related Classes of Distributions and Densities. viii, 173 pages, 1992.

Vol. 77: E. Mammen, When Does Bootstrap Work? Asymptotic Results and Simulations. vi, 196 pages, 1992.

Vol. 78: L. Fahrmeir, B. Francis, R. Gilchrist, G. Tutz (Eds.), Advances in GLIM and Statistical Modelling: Proceedings of the GLIM92 Conference and the 7th International Workshop on Statistical Modelling, Munich, 13-17 July 1992. ix, 225 pages, 1992.

Vol. 79: N. Schmitz, Optimal Sequentially Planned Decision Procedures. xii, 209 pages, 1992.

Vol. 80: M. Fligner, J. Verducci (Eds.), Probability Models and Statistical Analyses for Ranking Data. xxii, 306 pages, 1992. .

Vol. 81: P. Spirtes, C. Glymour, R. Scheines, Causation, Prediction, and Search. xxiii, 526 pages, 1993.

Vol. 82: A. Korostelev and A. Tsybakov, Minimax Theory of Image Reconstruction. xii, 268 pages, 1993.

Vol. 83: C. Gatsonis, J. Hodges, R. Kass, N. Singpurwalla (Editors), Case Studies in Bayesian Statistics. xii, 437 pages, 1993.

Vol. 84: S. Yamada, Pivotal Measures in Statistical Experiments and Sufficiency. vii, 129 pages, 1994.

Vol. 85: P. Doukhan, Mixing: Properties and Examples. xi, 142 pages, 1994.

Vol. 86: W. Vach, Logistic Regression with Missing Values in the Covariates. xi, 139 pages, 1994.

Vol. 87: J. Müller, Lectures on Random Voronoi Tessellations.vii, 134 pages, 1994.

Vol. 88: J. E. Kolassa, Series Approximation Methods in Statistics. Second Edition, ix, 183 pages, 1997.

Vol. 89: P. Cheeseman, R.W. Oldford (Editors), Selecting Models From Data: AI and Statistics IV. xii, 487 pages, 1994.

Vol. 90: A. Csenki, Dependability for Systems with a Partitioned State Space: Markov and Semi-Markov Theory and Computational Implementation. x, 241 pages, 1994.

Vol. 91: J.D. Malley, Statistical Applications of Jordan Algebras. viii, 101 pages, 1994.

Vol. 92: M. Eerola, Probabilistic Causality in Longitudinal Studies. vii, 133 pages, 1994.

Vol. 93: Bernard Van Cutsem (Editor), Classification and Dissimilarity Analysis. xiv, 238 pages, 1994.

Vol. 94: Jane F. Gentleman and G.A. Whitmore (Editors), Case Studies in Data Analysis. viii, 262 pages, 1994.

Vol. 95: Shelemyahu Zacks, Stochastic Visibility in Random Fields. x, 175 pages, 1994.

Vol. 96: Ibrahim Rahimov, Random Sums and Branching Stochastic Processes. viii, 195 pages, 1995.

Vol. 97: R. Szekli, Stochastic Ordering and Dependence in Applied Probability. viii, 194 pages, 1995.

Vol. 98: Philippe Barbe and Patrice Bertail, The Weighted Bootstrap. viii, 230 pages, 1995.

Vol. 99: C.C. Heyde (Editor), Branching Processes: Proceedings of the First World Congress. viii, 185 pages, 1995.

Vol. 100: Wlodzimierz Bryc, The Normal Distribution: Characterizations with Applications. viii, 139 pages, 1995.

Vol. 101: H.H. Andersen, M.Højbjerre, D. Sørensen, P.S.Eriksen, Linear and Graphical Models: for the Multivariate Complex Normal Distribution. x, 184 pages, 1995.

Vol. 102: A.M. Mathai, Serge B. Provost, Takesi Hayakawa, Bilinear Forms and Zonal Polynomials. x, 378 pages, 1995.

Vol. 103: Anestis Antoniadis and Georges Oppenheim (Editors), Wavelets and Statistics. vi, 411 pages, 1995.

Vol. 104: Gilg U.H. Seeber, Brian J. Francis, Reinhold Hatzinger, Gabriele Steckel-Berger (Editors), Statistical Modelling: 10th International Workshop, Innsbruck, July 10-14th 1995. x, 327 pages, 1995.

Vol. 105: Constantine Gatsonis, James S. Hodges, Robert E. Kass, Nozer D. Singpurwalla(Editors), Case Studies in Bayesian Statistics, Volume II. x, 354 pages, 1995.

Vol. 106: Harald Niederreiter, Peter Jau-Shyong Shiue (Editors), Monte Carlo and Quasi-Monte Carlo Methods in Scientific Computing. xiv, 372 pages, 1995.

Vol. 107: Masafumi Akahira, Kei Takeuchi, Non-Regular Statistical Estimation. vii, 183 pages, 1995.